W9-BLI-257

MANUAL OF

Fish Health

Everything you need to know about aquarium fish, their environment and disease prevention

Dr Chris Andrews, Adrian Exell & Dr Neville Carrington

Firefly Books

A FIREFLY BOOK

Published by Firefly Books Ltd. 2003

First printing

Publisher in Cataloguing-in-Publication Data (U.S.)
(Library of Congress Standards)

Andrews, Chris.
Manual of fish health : everything you need to know about aquarium fish, their environment and disease prevention / Chris Andrews ; Adrian Exell ; Neville Carrington. – 1st ed.
[208] p. ; col. photos. : cm.
Includes bibliographical references and index.
Summary: Health care for all types of fishes: freshwater and marine, tropical and temperate, pond and aquarium.
ISBN 1-55297-824-9
1. Marine aquarium fishes—Diseases. 2. Aquarium fishes—Diseases. I. Exell, Adrian. II. Carrington, Neville. III. Title.
639.34 21 SF458.5.A565 2003

National Library of Canada Cataloguing in Publication Data

Andrews, Chris, 1953-
Manual of fish health : everything you need to know about aquarium fish, their environment and disease prevention / Chris Andrews, Adrian Exell, Neville Carrington.
ISBN 1-55297-824-9
1. Aquarium fishes--Health. I. Exell, Adrian II. Carrington, Neville III. Title.
SF457.A53 2003 639.34 C2003-900786-3

THE AUTHORS

Dr. Chris Andrews is a fisheries scientist and consultant. He has travelled widely as an 'ambassador' for the fishkeeping hobby and enjoys a well-earned reputation for being a clear communicator, with several books, many magazine articles and regular television appears to his credit.

Adrian Exell is a fisheries scientist and diving enthusiast. He is a Development Manager for a well-known aquarium products company.

Dr. Neville Carrington is an acknowledged innovator in the design of aquarium products including water treatments and aquarium hardware. He has a Phd in Pharmaceutical Engineering Science and has extensively researched fish diseases and water chemistry.

Dr. Peter Burgess is a fish health consultant and university lecturer, specializing in ornamental fish. He holds degrees in parasitology, microbiology and fish biology. Dr. Burgess is a keen fish hobbyist and writes regularly for aquarium and koi magazines.

CONSULTANT

Peter W. Scott MSc., BVSc, MRCVS., MIBiol. is a member of the Zoo and Aquatic Veterinary Group, a practice that works solely with zoo and aquatic animals. He is the Veterinary Advisor to Ornamental Fish International, an accomplished author and an experienced lecturer on many aspects of fish health.

Above left: A broad stream rich in plant and animal life, including a range of freshwater fishes. The water quality in such a stream is a product of the rocks and soil over which it flows and any run-off from the surroundings fields. Since fishes have evolved to thrive in the conditions within their own environment, maintaining their health in the aquarium or pond is first and foremost a question of reproducing those conditions as closely as possible.

Credits
Editor: Geoff Rogers
Designer: Jill Coote
Color reproductions:
Fotograhcs Ltd. London – Hong Kong
Contemporary Lithoplates Ltd.
Filmset: Gee Graphics

Published in Canada in 2003 by
Firefly Books Ltd.
3680 Victoria Park Avenue
Toronto, Ontario, M2H 3K1

Published in the United States in 2003 by
Firefly Books (U.S.) Inc.
P.O. Box 1338, Ellicott Station
Buffalo, New York 14205

Printed in Italy

CONTENTS

HOW TO USE THIS BOOK

The chapters in this book are presented in a logical sequence that opens with the concept that good health is a delicate balance between many factors – not least of which is the influence of the environment – and concludes with a look at a representative range of remedies and treatment techniques. In a sense, the book is divided into 'positive' and 'negative' parts. Ill-health does gain a mention in Chapter 1, but the essential tone of the first four chapters is positive, with a discussion of fish anatomy and physiology in Chapter 2 being followed by an exploration of the often confusing intricacies of water chemistry in Chapter 3, while Chapter 4 offers practical advice on positive planning for good health. The book then pivots at Chapter 5, with the recognition of ill-health forming the prelude to the disease-orientated sections that follow in Chapter 6. The series of diagnostic charts in Chapter 5 are designed to provide a simple visual key to a range of common health problems in fish and form a useful 'springboard' into the detailed coverage of pests and diseases in Chapter 6. As the presentation opposite shows, using these charts is simply a matter of deciding which photograph and caption best describe the symptoms in your fish and then referring to the relevant pages in Chapter 6. Of course, not all disease conditions can be so conveniently presented, but this arrangement does at least provide a first step into the difficult territory of diagnosis and treatment. Although treatment and control measures are reflected in Chapter 6, it is vital to refer to the more detailed coverage in Chapter 7 before taking any action.

The diagnostic chart from Chapter 5 reprinted opposite shows a range of parasites that can be clearly seen on the skin and fins. Use the photographs and the accompanying captions as a guide to locating the relevant section in Chapter 6 that examines each particular parasite in more detail. We have carried one example through by tracing the leech diagnostic signs shown in the chart to the spread of Chapter 6 dealing with leeches and their control. The pests and diseases featured in Chapter 6 are presented in A-Z order of the most widely used common name. This spread on leeches shows the typical approach taken in the chapter, with the text organized under four headings:

- **Caused by**
- **Obvious symptoms/signs**
- **Occurrence of the disease/problem**
- **Treatment and control**

Where appropriate, photographs are included to show the disease organisms or pests, plus their effect on the fish. Life cycle diagrams and text panels provide additional information throughout.

Measurements and abbreviations

Length

m metre
cm centimetre (one hundredth of a metre)
mm millimetre (one thousandth of a metre)
micron micrometre/not abbreviated (one millionth of a metre or one thousandth of a millimetre)
nm nanometre (one thousand millionth of a metre or one millionth of a millimetre)

Conversion factors
metres to yards: ×1.094
metres to feet: ×3.281
cm to inches: ×0.3937
mm to inches: ×0.03937

Weight

kg kilogram (a thousand grams)
gm gram
mg milligram (one thousandth of a gram)
microgram not abbreviated (one millionth of a gram)
ppm parts per million (equivalent to mg/litre)

Conversion factors
kilograms to pounds: ×2.205
grams to ounces: ×0.3527

Volume
litre (not abbreviated)
ml millilitre (one thousandth of a litre)

Conversion factors
litres to Imp gallons: ×0.22
litres to US gallons: ×0.26

Temperature

°C Degrees Centigrade

Conversion factor
°C to °F: ÷ 0.555 + 32

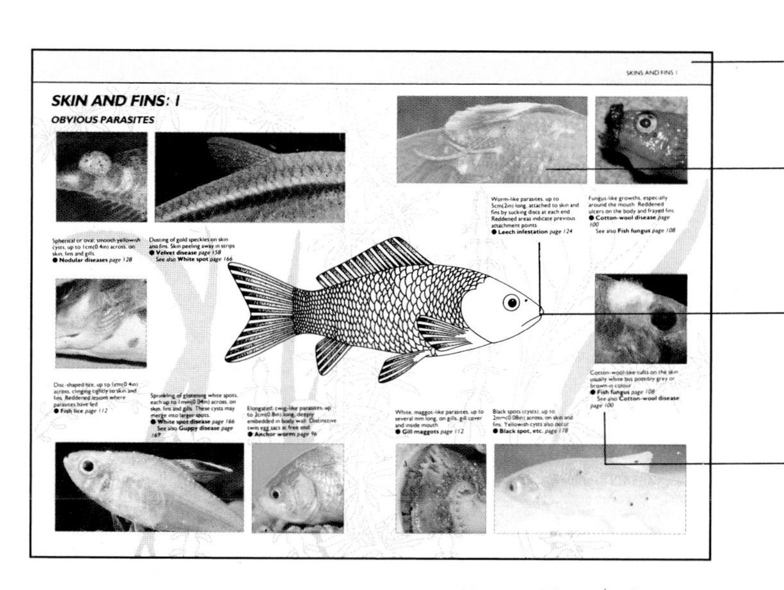

Colour banding clearly identifies each chapter.

The photographs in these diagnostic charts show the signs or symptoms of common health problems.

Each caption describes the symptoms, giving sizes of the disease organisms or parasites involved.

Page references to the relevant sections of Chapter 6 are quoted at the end of each caption.

The sections in Chapter 6 are arranged in A-Z order of the most widely used common names.

Parasite life cycle diagrams show larval stages, hosts and focus on control and treatment methods.

In the diagnosis charts and throughout Chapter 6, running heads show the theme of each spread.

LEECH INFESTATION

LEECH INFESTATION

Caused by
Piscicola geometra and various other leeches.

Obvious symptoms
Large leeches (up to 5cm/2in long) firmly attached to the skin, fins and, perhaps, gills. Heavily infested fish may appear listless, thin and occasionally behave in an agitated fashion. Reddened areas on the body indicate previous points of attachment, which may become infected with fungus. Leeches may, via their bloodsucking feeding habits, transmit microbial diseases between fish.

Occurrence of the problem
Leeches and other flatworms are often introduced into ponds and, more rarely, aquariums with new fish, plants (especially plants from local ponds or rivers) or live food. Not all flatworms are parasitic, however; some are simply scavengers.

Leeches reproduce by means of eggs, and these are laid in ponds in temperate regions anytime from spring to autumn. These eggs are quite resistant to treatment and, once they hatch out, the parasites can live for some time away from the fish host.

Treatment and control
Organophosphorus insecticides, such as metriphonate, may be used to treat leech infestations, although the active ingredients may not eliminate the eggs. Consequently, more than one treatment may be necessary. Note that some fish and many invertebrates are very sensitive to these chemicals. Leeches have also been removed from freshwater pond fish by placing the infested fish in a 2-3 percent solution of cooking salt (sodium chloride) for 15-30 minutes.

Leeches are most often a problem in ponds rather than in aquariums. One of the most effective ways of eradicating leeches from a pond is to drain and dry the pond for several days, discarding any plants and replacing them from a leech-free source. Adequate drying will kill both leeches and larvae within eggs.

Above: Typical damage to the skin of a fish caused by the bloodsucking feeding habits of leeches. Such lesions may become infected.

Below: Goldfish may suffer from heavy infestations with leeches as they emerge from their winter hibernation in garden ponds.

The life cycle of the fish leech (*Piscicola geometra*)

Adult leeches remain attached to the fish host for 2-3 days at a time, feeding on blood.

Leeches may leave the fish to digest their meal or to lay eggs.

Leeches lay eggs in dark brown oval cocoons attached to plants and rocks.

Leeches hatch from eggs and must find a fish to feed upon.

Treat infested fish with a saltwater bath or suitable remedy to eradicate leeches.

Control leeches and eggs by draining pond dry for several days and discarding plants.

WHAT IS A LEECH?

Leeches are classified in the phylum Annelida, together with other 'worms', such as earthworms, bristleworms, whiteworms, and *Tubifex* worms, etc. As a group, the annelids usually have muscular, well-organized bodies, with a well-developed nervous system and alimentary tract. Most are free living.

The leeches are placed in a separate grouping within the annelids – the Hirudinea. There are over 300 species of leeches. Most live in fresh water; a few live on land or in the sea. They usually have a sucker at both ends of the body, and these suckers are powerful attachment organs. Most leeches measure a few centimetres in length; some are considerably longer.

Some leeches are predatory, feeding on small invertebrates, but many are parasites that feed on the blood and tissue fluids of other invertebrates, as well as fish, amphibians and other vertebrates. Bloodsucking leeches have sharp, cutting mouthparts, and secrete an anticoagulant to prevent the blood from clotting. A single leech can, in a single meal, take in ten times its own weight of blood and then fast for several months.

Leeches are sometimes confused with turbellarian flatworms, although the latter are usually small, without obvious suckers, and move with a smooth gliding motion. Although leeches are hermaphrodite (i.e. contain male and female sexual organs), they cannot fertilize themselves, but need to cross-fertilize with another leech. Leeches may lay their eggs in cocoons above the water level. The eggs are resistant to many chemicals, but are usually susceptible to thorough drying.

Above: A fish leech, *Piscicola geometra*, engorged with blood after feeding. Up to 5cm(2in).

Where possible, colour photographs show the disease organism or parasite on the fish.

Photographs also show the effects of or the damage caused by the disease or parasite under review.

Text panels provide useful background information on the condition and/or disease in question.

CHAPTER I

THE BALANCE OF HEALTH

As a fishkeeper, you must accept a basic obligation to the totally dependent creatures in your care. Your first priority must be to provide the best possible conditions to safeguard their health and to promote their well-being. Accepting and discharging this responsibility to the full yields three fundamental benefits. First, it will minimize stress on the fish, which is unavoidable in their unnatural captive environment. Secondly, it will lead to relatively trouble-free fishkeeping. And thirdly, well-cared-for fishes will reward you by displaying their best colours and their most natural behaviour for you to enjoy.

In this opening section, we look briefly at the relationship between fish, their environment and the pathogens (i.e. the disease-causing organisms) that potentially threaten their health and survival.

Fish, pathogens and the environment

Since fish consist of 80 percent of the raw material that makes up the environment in which they live – namely water – and only a simple membrane separates the two, it is not surprising that fish are uniquely influenced by any alterations in the environment and that any fluctuations have a significant impact on their health.

Many potential fish pathogens are a constant and natural part of the environment, usually without causing disease problems and mortality. For example, fish usually carry small populations of protozoan parasites that feed mainly off surplus tissue and are kept under control by the fish's immune system. After all, it is not usually in a parasite's interest to kill its host. (There are some exceptions, however, where the parasite's life cycle involves killing the fish so that it is eaten by the next scavenging stage in the cycle.)

This unique relationship between fish, their pathogens and the total environment in which they live means that under normal environmental circumstances there is a balance between the fish and the pathogens. In this case, the fish's immune system keeps any problems under control. However, if there is an alteration in one or more of the environmental characteristics then there may be a shift in the balance to the benefit or detriment of either the fish or the pathogen. An example of an environmental change that runs counter to a pathogen is a fall in temperature that reduces the

Left: The health of any aquatic system, such as this community of freshwater tropical fishes, is held in a delicate balance. Maintaining this balance is the key to successful fishkeeping.

virulence of that pathogen, by slowing down its rate of multiplication or even halting its life cycle.

If the environmental shift is against the fish, then – in addition to any direct physiological impact – the fish usually becomes stressed, its immune system is suppressed and it becomes more susceptible to disease. In this situation a disease outbreak may occur, particularly if the environmental shift also favours the pathogen. As an example *Flexibacter* bacteria are commonly present in fish-holding water, but the fish's immune system keeps them at bay. However, when excess food rots in the environment the bacterial population rises and, as the bacteria break the food down, they produce excess amounts of ammonia. Ammonia is a toxin that irritates the fish's gills, causing them to overproduce a layer of cells on the delicate gill surface. Since these cells are more prone to bacterial infection, the combined effect of these circumstances is an episode of bacterial gill disease in the fish.

Another relevant disease-causing factor is the introduction of disease organisms into the environment on fish, plants or decorations. This can result in the introduction of some types of pathogen that, as a consequence of their action, inevitably kill the fish host. The introduction of different types or strains of pathogens will also upset the natural immunobiological balance of a community of fishes. The indigenous fishes will have developed a degree of immunity to most of the disease organisms in their environment but may not have encountered the new strains of pathogen being brought in. On the other hand, the newly introduced fishes will also have to develop an immunity to the population of pathogens in their new environment. Therefore, introducing new fishes greatly increases the chances of a disease outbreak occurring. This underlines the importance of careful quarantining, a simple precaution that reduces the likelihood of disease when you are building up a fish population from different sources. (For guidance on quarantine, see pages 70-73.)

Stress and disease

Stress is a vital factor in fish health. The varied forms it takes and its effects on fish have been investigated by biologists and fisheries scientists in both wild and captive fish populations. Here, we draw some simple conclusions from what is a highly complex story.

Factors that have a negative impact on fishes, such as handling, overcrowding, poor environmental conditions and unsuitable or aggressive tankmates, are called 'stressors'. The stress response these stressors cause is defined as the sum of the physiological responses the fish makes to maintain or regain its normal balance. Some stress responses are common to all stressors; others are specific to one particular type.

The most basic stress response is to escape from imposed danger, which may take the form of a natural predator or, in captivity, the fishkeeper attempting to catch the fish in a net for closer examination or transfer it to another tank or pond. The first part of this response is the preparation of the body for escape, which, in biological terms, involves the release of hormones that channel all the fish's energy to power the locomotory muscles. Unfortunately, this alarm response has long-term detrimental effects. For example,

Environment and disease

Above: Poor conditions, such as in this algae-ridden tank, can make fishes more open to infection.

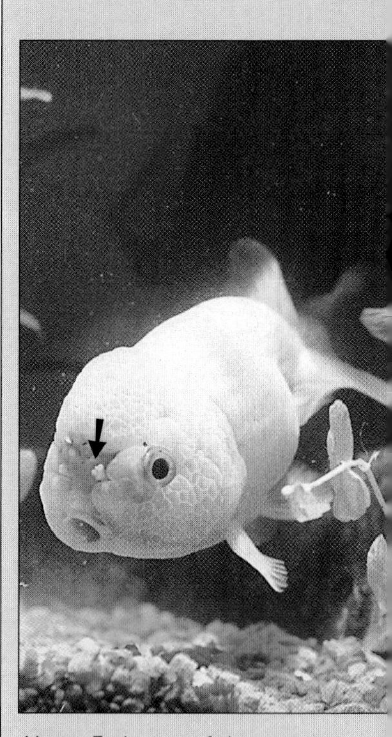

Above: Early signs of ulcer disease can be seen on the snout of one of these white lionhead goldfishes. The bacteria responsible are usually present as low-level infections ready to erupt if the fish is stressed or kept in unsuitable conditions.

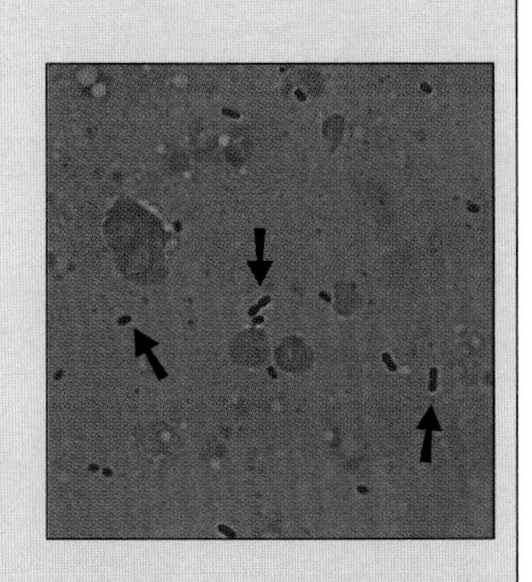

Above: Bacteria, such as these *Aeromonas* sp. that cause ulcer disease, abound in most aquatic environments.

one of the hormones released is adrenalin, which apart from 'quickening' the body for instant action also disturbs the osmoregulation (i.e. control of the salt/water balance) within the fish. Another hormone involved, cortisol, affects the white blood cells and reduces their effectiveness in the immune system. The second part of the response involves the recovery of the fish's equilibrium. Thus, the alarm response is clearly a compromise between the short-term need to reach immediate safety and the longer term side-effects of the physiological changes involved. It can be an uneven compromise; a fish that undergoes an alarm response as a result of stress applied for even a short period of time can take hours or even days to recover its equilibrium.

The stress response in fish is generally considered to be less well suited to reacting to chronic environmental stressors. This is because fish have evolved in a relatively stable natural environment and have not, therefore, developed an effective system to deal with environmental changes or chronic stressors. In the case of a negative environmental change, a fish's first response is also an alarm stage, in which it attempts to get away from the problem. If escape is not possible, then it is followed by an adaptive stage during which the fish's body attempts to react to the environmental change. Initially, the fish's physiological compensation tends to over-react. Then, over a longer term, it returns to a new equilibrium in which the fish reaches optimum adaptation, both physiologically and behaviourally, to survive the new environmental conditions. During the adaptive stage of the stress response, the fish channels much of its resources into dealing with the stressor. As a result, its immune system functions less effectively and hence the fish is more prone to disease problems.

Although a fish may successfully adapt to new conditions, its performance in terms of growth, breeding and disease immunity may be reduced. This adaptive stage may last from four to six weeks. If a fish is constantly exposed to stressors such as a steady deterioration of environmental conditions or continual bullying by tank-mates the adaptive stress response is likely to be so extended, and the fish's normal functions so disturbed, that its chances of survival are significantly reduced. If the environmental change is so great that the fish cannot compensate – then the fish's stress response finally reaches a fatal exhaustion stage.

Although stressors and the fish's stress response can result in the fish succumbing to disease because of the immune system's reduced effectiveness, the diseases themselves can also be considered as stressors in their own right. In this case, the fish's adaptive response is an attempt to counter the disease effects.

Even from this simple introduction it is clear that minimizing stress is an extremely important part of good fishkeeping. How to reduce it at every possible stage is a recurring theme in all the fishkeeping advice given throughout this book. Without doubt, careful forethought and planning, and effective management of water quality and fish populations, are the key elements in successful trouble-free fishkeeping. As with many health-related endeavours, the old adage is consistently true: prevention is far better than a cure.

CHAPTER 2

LIVING IN WATER

Water covers just over 70 percent of the Earth's surface and is home to some 24,000 species of fish in an incredible variety of shapes, sizes and colours. These have evolved over millions of years to fill every conceivable niche in the world's watercourses and seas, each species supremely suited in form and function to its own particular aquatic existence.

Although water is the source of life, it is, nevertheless, a highly demanding environment in which to live. In this chapter we look briefly at how fishes are adapted to live in water and how they relate to their environment. Armed with a basic knowledge of their anatomy and physiology, you should find it easier to understand the fishes in your care and to provide the conditions that will promote their health.

Basic body shape

The external body form of a fish is a function of the environment in which it evolved and how it lives. How active it is, for example, how it feeds, whether it is a predator or simply a potential prey and thus what attack or defensive systems it needs to survive.

Within the bony fishes, the most highly 'developed' group, every possible specialization of exterior form has evolved, allowing species of fish to fully exploit the entire aquatic environment. Among aquarium fish, two examples serve to illustrate this diversity. The common plec (*Hypostomus plecostomus*), for example, has a mottled brown laterally compressed body and a ventral sucker mouth that are ideal for shuffling along rocks and grazing their algal cover as it keeps station in the raging torrents of South American rivers. In complete contrast, the zebra danio (*Brachydanio rerio*) has a streamlined body adapted for continuous fast movement in the upper water layers, its upturned mouth enabling it to sip insects delicately from the surface film.

Moving through water

Since water is 800 times denser than air, moving through it is fraught with problems, such as drag, negative buoyancy (sinking) and the sheer effort required to displace such a dense medium. Thus, body form and locomotory adaptations are closely related to how a fish lives and its need for movement.

Left: A shoal of Australian rainbowfish (*Melanotaenia boesemani*), hardy fishes that are well adapted in their shape and physiology for their active, midwater-swimming life style.

Here, three contrasting examples illustrate the wide variety in fish form. The mbuna cichlids of Lake Malawi have a life style that requires optimum manoeuvrability and the ability to hover in one place. Their need to overcome negative buoyancy has led to the development of an adjustable air sac, called a swimbladder, that enables them to 'hang' at any level in the water (in fact, most fish possess a swimbladder). Thus stabilized, they use their well-developed paired fins – the pectorals and pelvics – as effective steering mechanisms. As a general trade-off, these fishes tend not to have very streamlined bodies, and swimming speeds are therefore limited by drag. These fishes typically have two types of muscle: brown muscle, which can be used for continuous activity because it has a good blood circulation and is constantly supplied with oxygen, and a large proportion of powerful, so-called 'anaerobic' white muscle, which can be used only for short emergency bursts of speed because it quickly builds up an 'oxygen debt' (see the section on metabolism, page 24).

Continuously swimming midwater fish, such as tuna and mackerel, need optimum streamlining features to minimize energy wastage by reducing drag and assisting displacement. These species tend not to have swimbladders because the increase in cross-sectional area and the higher drag factor that the swimbadder would introduce would offset any advantage gained by reducing the tendency to sink. Muscle is primarily brown for constant swimming. The fins are used only for turning and are usually held against the body during swimming.

Sedentary bottom-dwelling fish, such as the suckermouth and whiptail catfishes, have minimal locomotory requirements. These fish tend to be dorso-ventrally compressed and obviously have no need for a swimbladder. In such fishes, adaptations of form geared towards feeding, camouflage and defence take precedence over those for rapid midwater swimming.

Below: The giant whiptail catfish (*Sturisoma panamense*), from the rivers of Colombia, has an ideal flattened body shape for its bottom-dwelling way of life.

Above: The marine long-nosed filefish (*Oxymonocanthus longirostris*) typically adopts this head-standing pose as it uses its pectoral fins to manoeuvre in search for food.

Below: This cichlid, *Pseudotropheus lombardoi*, has a swimbladder and well-developed pectoral fins to enable it to hover as it feeds on algal growths in Lake Malawi.

Above: Camouflage is the primary aim of the body shape in the South American leaf-fish (*Monocirrhus polyacanthus*), as it waits to ambush its unsuspecting fish prey.

Vital senses

Like all animals, fish need to be aware of what is happening around them; they need sensory equipment to carry out and coordinate such vital functions of life as navigation, communication, attack, defence and food location.

As a sensory environment, water differs from air in a number of important ways:

- Light is rapidly absorbed by water and turbidity further reduces visibility
- Sound as pressure waves travels faster and further in water than in air, simply because it is a denser medium
- Since all chemicals, including those emanating from food substances, are dissolved in water and readily diffuse through it, the senses of smell and taste hold a special significance in the aquatic environment
- the water surrounding fish is also an electrolytic solution, (i.e. able to conduct electricity because molecules in solution exist as electrically charged ions), hence some fish are able to use the unique sensory tool of electroreception.

Sight

Fishes' eyes are very similar in structure to those of other vertebrate animals, the main adaptation being the almost spherical shape of the lens. Fish can focus selectively on both near and far away objects, and their field of view is generally determined by the position of the eyes on the head. In fishes such as neon tetras (*Paracheirodon innesi*), for example, the eyes are located at the sides of the head and provide good 'defence' vision in a wide arc on each flank of the fish. In pike cichlids (*Crenicihla* species), however, the eyes are positioned further forward and enable the fish to focus more clearly on its intended prey ahead. Except for a few species with modified eyes, fishes' vision is limited above the water due to the distortion of light rays at the water surface. Detailed examination of the eye structure suggests that fish see their world in much the same way as we do – in glorious 'technicolour'.

Basic fish anatomy

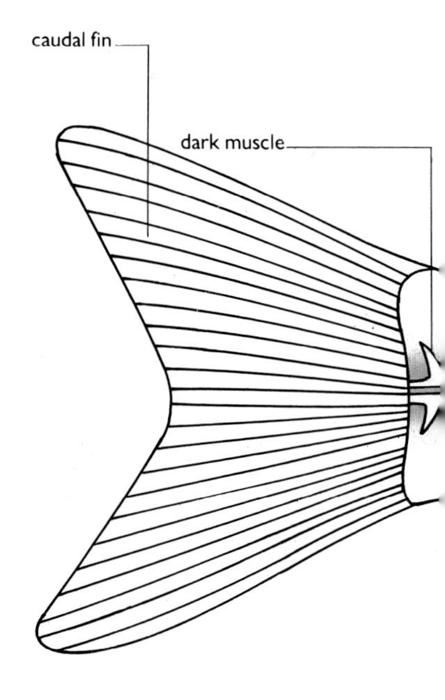

Right: Forward-facing eyes enable this predator, *Boulengerella lateristriga*, to focus accurately and judge the range of its prey.

Eye position and field of view

Predatory fishes

Non-predatory fishes

High-definition hunting sight

Low-definition warning/alert sight

Left: These two drawings compare the fields of view and areas of sharp focus that result from different positioning of the eyes. In non-predatory fishes, the eyes are usually located at the sides of the head and provide wide angle 'warning' vision of impending danger. In predators, the eyes are usually further forward to provide a larger area of high definition 'hunting' sight in sharp focus directly in front of the fish.

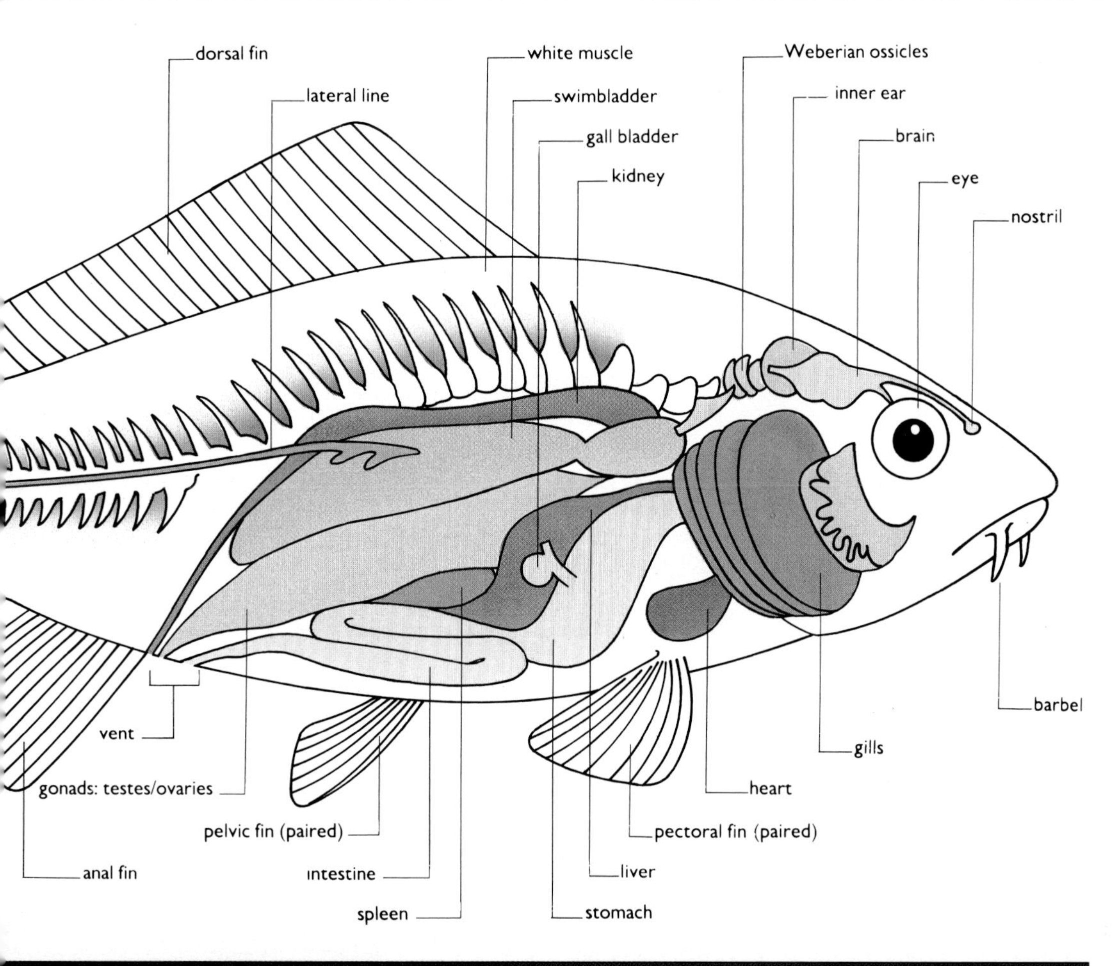
dorsal fin
lateral line
white muscle
swimbladder
gall bladder
kidney
Weberian ossicles
inner ear
brain
eye
nostril
barbel
gills
heart
pectoral fin (paired)
liver
stomach
spleen
intestine
pelvic fin (paired)
anal fin
gonads: testes/ovaries
vent

Above: The blind cavefish (*Astyanax mexicanus*) does not need functional eyes in the subterranean streams in which it lives. It relies on a highly developed lateral line plus other orientation receptors and chemoreceptors.

Sound and pressure waves

Fish rely heavily on the sensory reception of sound, which in water manifests itself as pressure waves. They have a very sensitive so-called 'lateral line' system that consists of a series of canals and pits set just below the skin, the main canal running along the midline of each flank. This is tuned to ignore background noise and picks up unusual low frequency sounds vibrating at 1/10 to 200 hertz (cycles per second). Fish also have an inner ear, which picks up higher frequency sound (up to 8000 hertz). In some fish species, such as carp, the inner ear system is highly developed, with the swimbladder acting as a receiver and amplifier for sounds that are passed to the inner ear by means of a series of connecting bones called the Weberian ossicles.

Orientation

Sense receptors within the inner ear and in closely related structures also enable fish to orientate themselves in three dimensions within the aquatic environment. The otoliths (ear bones) register 'tilt' of the head and respond to linear acceleration, while the movement of fluid within the semicircular canals triggers receptors that register turning.

Smell and taste

Responding to chemicals diffusing through the water is a highly specialized sense in fishes, since it is vital for communication and discovering food. The distinction between 'smell' and 'taste' is difficult to make; both senses are best described simply as 'chemoreception'. There are special chemoreceptor sites concentrated in the nasal openings, scattered in the mouth, around the head and, in some species, even over the body. In catfishes and loaches, chemoreceptors and touch sensors are concentrated on barbels, or 'whiskers', around the mouth. These are generally used for finding food in low light and many of these fish are nocturnal.

Electroreception

Some fish, such as the elephant-nose of the Mormyrid family, use the lateral line as an electrosensory system. Electric pulses emitted by an organ near the tail set up an electric field in the water around the fish and sensory receptors near the head are able to detect minute changes within this field caused by the approach of other fishes or solid obstructions nearby. This system is used for navigation and communication and for finding food in low light conditions.

Coordination and control systems

The coordination and control of bodily processes in response to both external and internal stimuli is achieved by the brain in cooperation with the nervous and endocrine systems.

The brain receives and assimilates information from the sensory organs, and then coordinates and stimulates the correct responses from the appropriate organs within the body. The brain also integrates reflex actions, such as 'breathing' and heart function, and is the site of learning and memory.

The nervous system is responsible for rapid changes in physiological function. Nerve messages are electric pulses that travel very rapidly along the networks of insulated nerve fibres. Sensory nerves carry messages to the brain, while motor nerves carry messages from the brain to the response organ.

The endocrine system takes longer to respond and controls the vital organs to maintain a constant internal environment. The system consists of a number of endocrine organs that produce messenger chemicals, called hormones, that are carried in the bloodstream to their target organs.

Below: The shovel-nosed catfish (*Sorubim lima*) using its long barbels to probe its surroundings. The barbels are rich in touch sensors and chemoreceptors that relay environmental information to the fish's brain.

Blood circulation within a fish

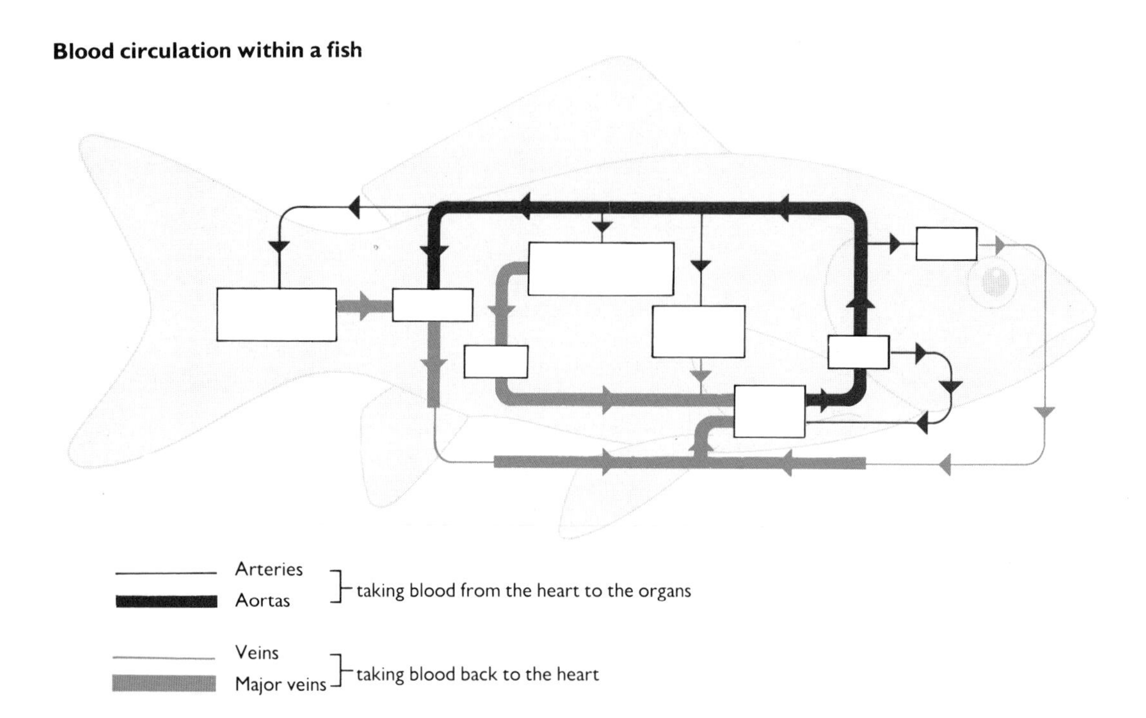

The cardiovascular system

The cardiovascular system forms an internal transport system connecting every organ and cell together and serving a host of functions. It consists of a pump – the heart – which circulates the carrying fluid medium – the blood – around an extensive pipework system of arteries, veins and capillaries. In fishes, the heart is a simple four-chambered structure with only two valves. It is not very powerful, which means that the circulation is slow and that the tissues farthest from the heart have to function inefficiently due to relatively low levels of food and oxygen and a build-up of waste products. The rate of circulation can be increased in response to environmental demands; hormones released by the endocrine system stimulate the heart to beat faster and thus transport larger volumes of blood. The smaller arteries can also dilate to reduce resistance to the blood flow.

Blood is a complex medium, 30-50 percent of which consists of blood cells. Most of these are red blood cells that carry oxygen, while the remainder are white blood cells associated with the immune system (see page 30). The rest of the vascular fluid – the plasma – consists mainly of water, salts and matter being carried, such as glucose for energy and waste products being 'collected' from around the body.

Nutrition

Fishes have evolved to exploit many different food sources. There are species which are solely carnivorous (i.e. meat-eating) and vegetarian species that graze aquatic plant matter, while many species are omnivorous, eating a combined diet. Individual species have evolved specialized mouth structures for coping with specific

Right: Quite clearly a carnivore, *Xenentodon cancila*, has long 'toothy' jaws to capture and swallow prey whole for digestion in the acid juices of its stomach.

Below: A royal plec (*Panaque nigrolineatus*) rasping algae from an aquarium rock. Its mouth and digestive system are adapted to cope with a vegetarian diet.

food sources and digestive systems tuned to deal efficiently with their diet. Vegetarian fish, for example, generally have a relatively long gut without a stomach, which allows the enzymes to be in contact with the food over an extended period so that they can work on the more difficult to digest vegetable matter. In contrast, carnivores have a much shorter gut with a stomach, in which food is retained in a very acid environment to encourage protein digestion. Although diet and digestive systems vary from fish to fish, the basic process of nutrition is essentially the same. Food is ingested and passes into the stomach, where digestion begins. This involves enzymes breaking the food down into its component parts, a process that continues as the food passes down the gut. In the lower gut the useful food materials are absorbed into the bloodstream and transported to where they are needed. The material remaining in the gut is voided as faeces. In general, 80 percent of food ingested is used and 20 percent is voided as waste.

All food consists of a combination of proteins, carbohydrates, lipids (fats), minerals and vitamins, and a careful balance of these components is essential in a fish's diet.

Proteins are made up of different combinations of 21 amino acids. These are the 'building blocks' of body tissue and so proteins are normally used in growth or tissue maintenance. Fish also have the ability to break amino acids down to produce energy; this occurs if the amino acids are excess to requirements, or if there is no other source of energy available. The breakdown process results in the production of a toxic waste product called ammonia.

Carbohydrates consist of long chains of simple sugars and are mainly provided by vegetable matter. The digestive process breaks carbohydrates down into a simple sugar called glucose for absorption. Glucose is either used immediately to produce energy in the process of respiration, or built into a product called glycogen, which is stored in the liver and muscles to provide energy later on.

Lipids are made up of fatty acid chains and are broken down into these during digestion. Once absorbed into the body, fatty acids are usually stored as fat deposits until required. When 'mobilized', the lipids are either oxidized in brown muscle to provide energy or converted to phospholipids, complex organic compounds used in the construction of vital cell structures.

Vitamins and minerals are essential for good health, for both incorporation into body structure and use in metabolic processes.

Metabolism

Metabolism is the collective term for all the chemical processes which give life to a fish. These processes use products called metabolites, which include organic food material and inorganic matter such as oxygen. There are two metabolic processes: catabolism, which breaks down metabolites to produce energy for activity; and anabolism, which uses metabolites to build new tissue for growth, reproduction and renewal of old tissue.

Metabolism is linked with all other body processes, either by providing energy to power them or by building and maintaining the structures necessary for them to function. Metabolism depends on nutrition and respiration for the provision of metabolites, on osmoregulation to provide a stable working environment, and on excretion to remove the useless and poisonous resultant waste products. (We shall be looking at all these essential processes in more detail later in this section.)

The rate of metabolism is controlled by hormones and is influenced by a number of factors, including prevailing environmental conditions (such as temperature, salinity and oxygen levels), the level of the fish's activity, the size of the fish (larger fish have a lower metabolic rate per unit weight), the fish's age (young fish are undergoing more growth than adults but have less reproductive requirements), and the fish's condition (fish in poor condition need to constantly replace damaged tissue).

Generally, only the surplus energy remaining after the demands of a fish's life style, including the energy used to combat stress and disease, is used for growth and reproduction. Therefore, good growth and active breeding behaviour can be taken as a fair indication of good living conditions.

Above: Koi eagerly take food pellets at the water surface. A quality processed food (e.g. flakes, pellets) will contain all the essential ingredients in the correct proportions.

Right: This flow chart outlines the main processes of nutrition and metabolism that occur in a typical fish. Digestion in the gut breaks down proteins, lipids (fats) and carbohydrates into amino acids, fatty acids and glucose respectively, which are foodstuffs in their most 'mobilized' form. All three materials can be either 'stored' or 'consumed' to provide energy. The 'stored' form of amino acids can be regarded as the proteins they are built into for growth and tissue repair. Lipids are 'stored' as fat deposits and/or as vital cell components. Glucose is 'stored' as glycogen in the liver and muscles. Glycogen can be reconverted to glucose, either in the presence of oxygen or, in emergency situations, by a 'short cut' process called glycolysis (shown at the right-hand side) that does not require oxygen until *after* the energy has been liberated. The main energy cycle at the bottom of the chart shows how glucose is oxidized to produce energy, with the accompanying waste products of carbon dioxide and water. In most bony fishes, the main nitrogenous waste product is ammonia (produced from the breakdown of amino acids) and this is excreted mostly through the gills.

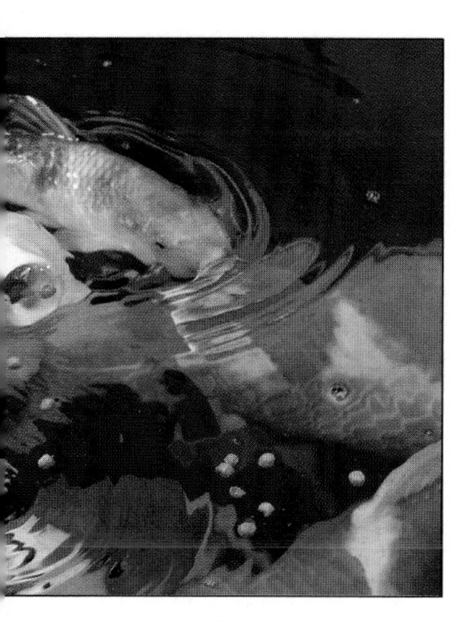

Under normal circumstances, energy is produced by a process of oxidation, which requires a constant and sufficient supply of oxygen. In emergency situations, energy can be produced rapidly in the white muscle by a process known as glycolysis, in which the hormone adrenalin causes glycogen to be converted to glucose and energy without requiring oxygen. During this process, lactate is produced. Since lactate is poisonous at high concentrations, however, glycolysis can be sustained only for a short period of time. The lactate accumulated is a kind of 'oxygen debt' because, eventually, oxygen and energy are required to break it down.

Waste products are generated during both energy production and in the constant maintenance and renewal of old body tissue with new. These waste products consist mainly of carbon dioxide, water, ammonia and some larger molecules, such as purine. All these waste products are toxic and need to be excreted. Carbon dioxide and ammonia are removed through the gills by diffusion, while water and purine (eventually urea) are removed by the kidney.

Nutrition and metabolism in fish

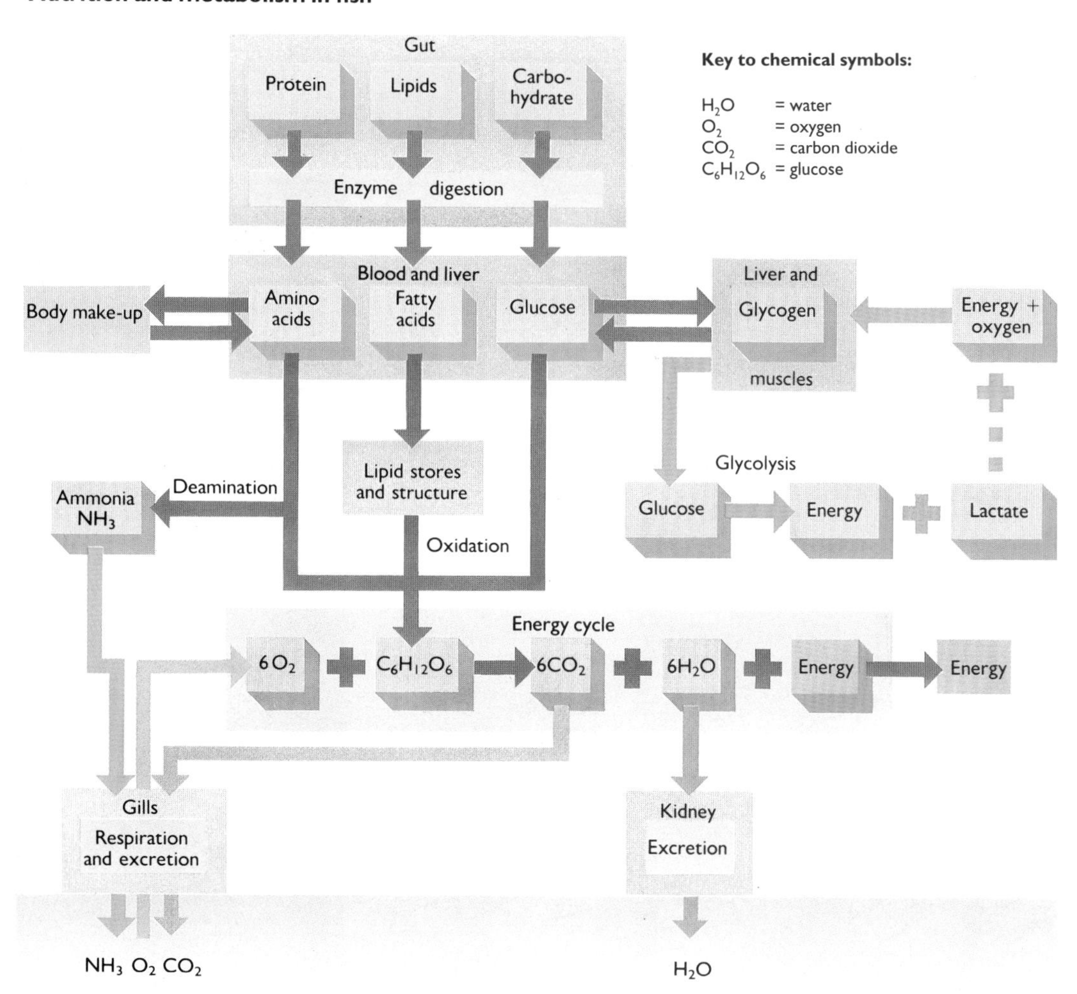

Osmoregulation – controlling internal salt/water balance

Fish are literally parcels of fluid within a fluid environment. In both marine and freshwater fish there is a difference between the salt concentration of the environment and their body fluids. Since the two are separated in places by very thin membranes, notably in the gills, it is not surprising that there is a constant tendency for salt or water to flow in or out of a fish's body. The processes at work here are diffusion and osmosis. If two solutions of different concentration are separated by a semi-permeable membrane, such as those that form the 'biological boundaries' of a fish, the ions of salt will move by diffusion through the membrane from the more concentrated to the weaker solution while the water molecules will move in the opposite direction by osmosis to dilute the stronger solution. As in many natural processes, there is a tendency towards equilibrium on both sides. For a fish's body to work efficiently, it is essential that it maintains its internal salt/water balance at a constant level, in spite of the salt concentration of the water in which it lives. Fish counteract the natural forces of diffusion and osmosis by means of a process called osmoregulation. Here, we see how the priorities differ in marine and freshwater fishes.

Above: A coastal lagoon in Sri Lanka, the typical brackish-water habitat of the scat (shown below) and other adaptable species.

Osmoregulation in marine fishes

Since sea water has a higher concentration of salts than the body fluids of marine fishes, there is a constant tendency for water to be lost from the fish's tissues and salt to flow in. Marine fish solve the dehydration problem by drinking vast amounts of water and excreting little urine. The influx of salt is counteracted by not absorbing salts from the sea water they drink and using energy to actively eliminate salt through special cells (known as chloride cells) in the gills.

Osmoregulation in marine fishes

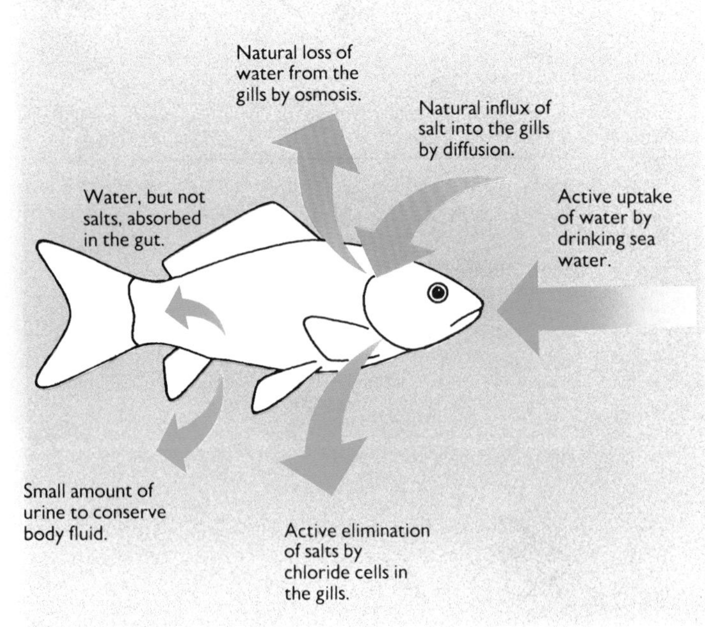

Below: A scat (*Scatophagus argus*), which has the ability to adapt its osmoregulatory system to cope with fresh, brackish or sea water.

Osmoregulation in freshwater fishes

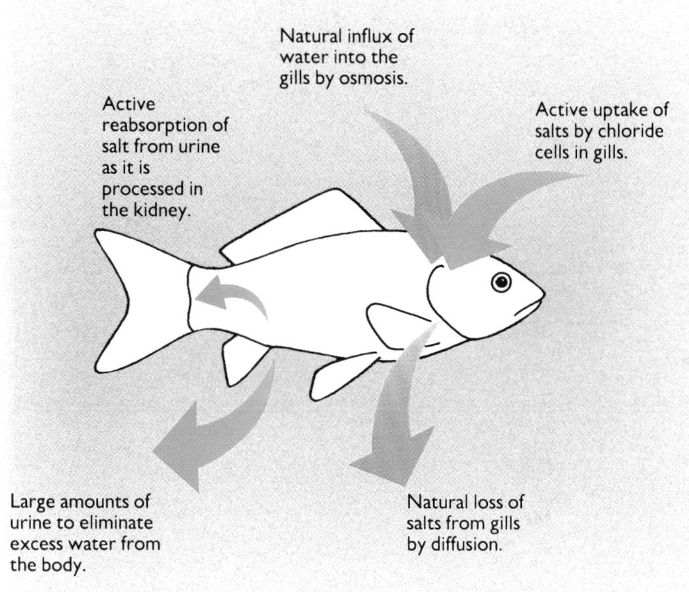

Osmoregulation in freshwater fishes

The situation in freshwater fishes is the opposite of that encountered by marine fishes, since the body fluid of freshwater fishes has a higher concentration of salt than the surrounding environment. The tendency, therefore, is for water to flow in and salt to be lost from the tissues. To counteract the first process, freshwater fish have very efficient kidneys which excrete water very rapidly. Salt loss is minimized by the efficient reabsorption of salt from urine before it is excreted and active uptake of salt through chloride cells in the gills. (How fish react to unsuitable osmotic conditions and how the osmoregulatory processes of different fish species vary with different environmental conditions are discussed on pages 45-46 and 49-50.)

Respiration

Fish require oxygen for life. The vital process by which they remove it from their aquatic environment and transfer it to their cells is called respiration.

Since water contains only five percent of the oxygen present in air, a fish's respiration system needs to be very efficient. It is essential to move large volumes of relatively oxygen-deficient water over the absorption surfaces to allow sufficient oxygen to be taken up. This transport mechanism also has to be energy efficient because, as we have seen, water is 800 times denser than air. To generate the necessary water flow, fish use the structure of the mouth, or buccal cavity, plus the gill covers and their openings, the opercula, to produce a very effective low-power pump. This produces a constant flow of water over the gas absorption surfaces of specialized respiration structures called gills.

Pumping cycle in fish respiration

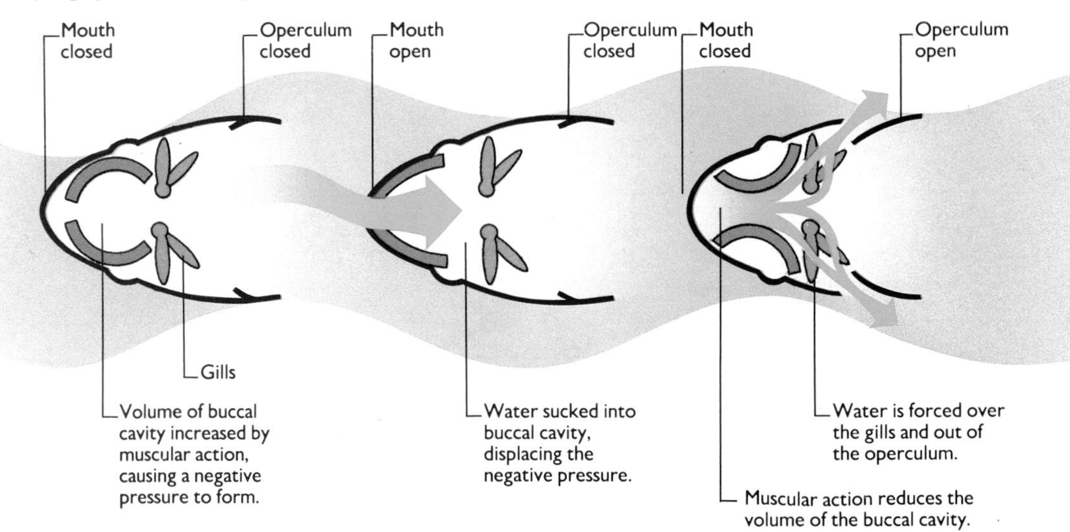

To absorb oxygen efficiently, the gill structures need to present a large surface area and a thin wall between the oxygen-carrying water and the blood. These parameters are limited by the fact that the more ideal they become for gaseous exchange, the more likely they are to promote osmoregulatory problems, since they provide an ideal site for water influx or loss. Gill structure therefore represents a compromise between the requirements for respiration and osmoregulation.

Oxygen is absorbed into the blood by simple diffusion. The blood flowing into the gills has a lower oxygen concentration than the surrounding water, so oxygen moves into the blood to redress the imbalance. This process is further improved by the fact that the blood is pumped in the opposite direction to the water moving over the gills. This countercurrent system ensures that the blood oxygen level remains below that of the water right across the gill and allows many fish to remove up to 80 percent of the water's oxygen content. The oxygen is actively picked up by haemoglobin in the red blood cells and transported to the body tissues, where a relatively high carbon dioxide level causes the oxygen to be given up for use in the cells' essential functions.

Above: This close view of fish gills clearly shows the bony gill arches and the rows of long gill filaments attached to them.

The countercurrent system in a fish gill

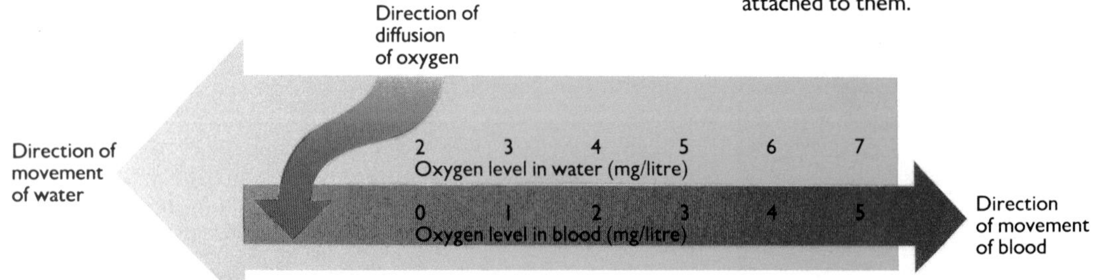

Left: These drawings show the sequence of events that enable fish to pump water over the gills. The sequence is really a cycle, with the third stage being followed by the first stage, and so on.

Carbon dioxide is a waste product of metabolism, but since this is soluble in the blood, it poses no problems in removal, finally diffusing out easily through the gill walls. Sometimes, carbon dioxide is transported in the blood as bicarbonate ions. In this case, it is traded at the gills for chloride salts as part of the osmoregulatory process. (See pages 53-55 in Chapter 3 for details of how fishes adapt to different environmental oxygen levels and to unsuitable oxygen and carbon dioxide levels in the water.)

Above: An even closer look at healthy gills shows the secondary lamellae, each supplied with blood capillaries for gaseous exchange.

Primary lamellae

Secondary lamellae

Structure of a fish gill

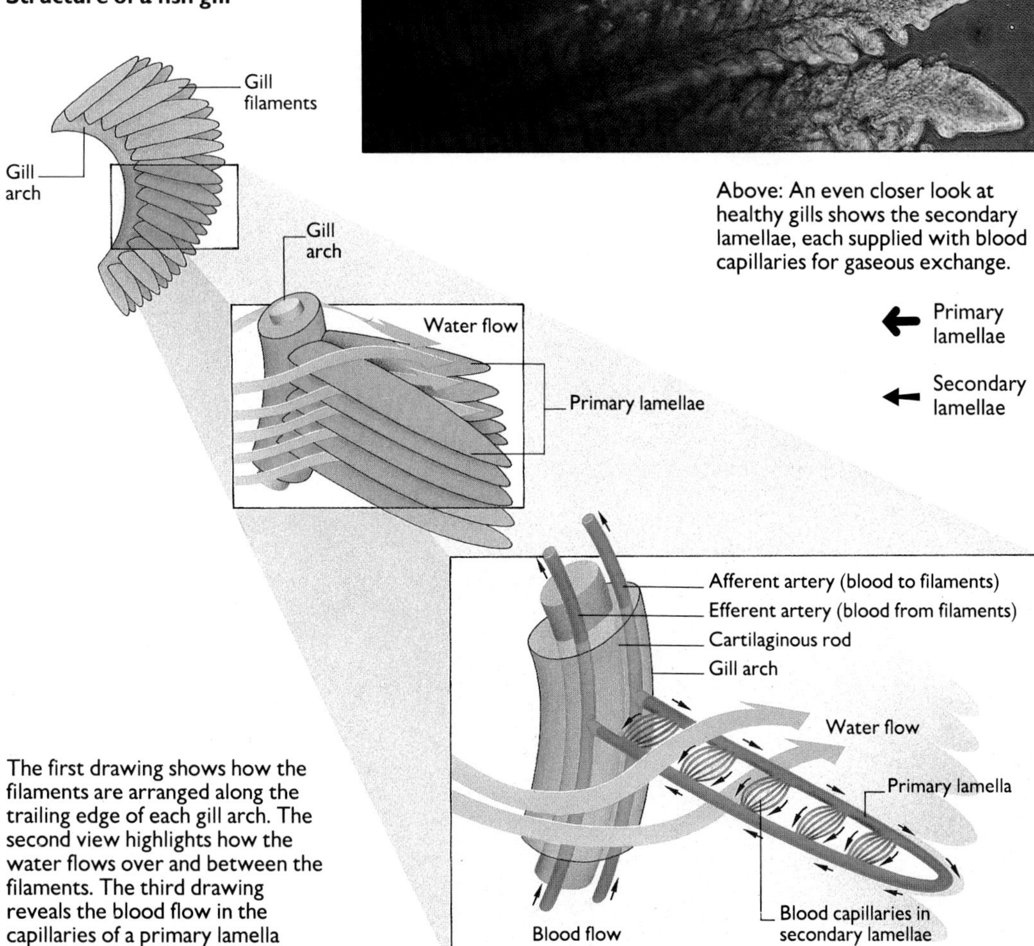

The first drawing shows how the filaments are arranged along the trailing edge of each gill arch. The second view highlights how the water flows over and between the filaments. The third drawing reveals the blood flow in the capillaries of a primary lamella where gaseous exchange occurs.

Reproduction

Survival of a fish species clearly depends upon its ability to reproduce itself, and fishes have developed a wide variety of successful strategies. All reproductive strategies are different means of applying the metabolic energy allocated to producing offspring in the most efficient way, i.e. simply in terms of balancing the number and size of eggs or young produced with the amount of effort applied in parental care. Some fish reproduce only when there is sufficient surplus energy, while others will spawn in direct proportion to the energy available. Some fish spawn irrespective of other requirements and may forfeit their life in doing so.

In bony fishes, the reproductive organs, or gonads, form into testes (male) or ovaries (female) according to the sex 'instructions' imprinted on the genes, although some fish have an amazing ability to change sex according to the prevailing conditions.

The immune system

Although not as advanced as that of mammals, fish do have an immune system that protects their bodies against disease. The first line of defence involves preventing disease organisms physically invading the body. Fish have an effective outer barrier in the form

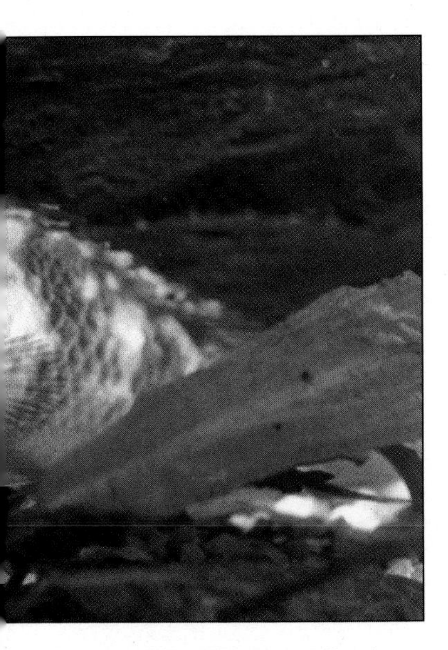

Above: This African bumphead cichlid, *Steatocranus casuarius*, shows clear signs of fungus infection following damage received in a fight. Any break in the protective layer of scales and skin lays a fish open to secondary bacterial or fungal infection. If the fish is in a weakened condition or being kept in poor water conditions, such infections may overwhelm the fish's immune system and eventually kill it. Prompt treatment with proprietary remedies is vital to keep such infections in check.

Left: A five-spot tilapia (*Tilapia mariae*) guarding its fry with parental concern. Fish show varying degrees of parental care for their eggs and young, from complete indifference to zealous sheltering over a period of several months. The cichlids, which include tilapias, are celebrated among aquarists for their range of spawning and fry-raising strategies.

of scales and the layers of the dermis and epidermis, all of which provide some protection against disease organisms and physical damage. This outer barrier is further improved by a covering of mucus that contains substances possessing anti-bacteria and anti-fungus properties. The mucous membrane is constantly being renewed, which also has the effect of sloughing off debris and dissuading the proliferation of external parasites. The other possible area of infiltration is through the digestive tract; here enzyme action and an unsuitable pH produce a hostile environment to discourage most pathogens.

If one of these barriers breaks down then pathogens can gain entry. This can occur through skin wounds and also through the gut; in some cases of stress, the gut seizes up and anaerobic fermentation and enzyme action attack the gut wall, allowing disease organisms to enter the tissues and bloodstream.

There are a number of substances in the bloodstream that give some general immunity by immediately attacking any pathogens. These include the antiviral chemical interferon and C-reactive protein, which acts against bacteria and viruses. The body's first coordinated response to foreign bodies in the bloodstream is to seal off the entry site to prevent osmoregulatory problems and hamper the spread of the disease organism throughout the body. This inflammation reaction is achieved by histamines and other products released by damaged cells that cause blood vessels in the area to close up. At the same time, the blood protein fibrinogen and clotting factors seal the area off with a physical barrier of fibrin. White blood cells are attracted into the area and these ingest the foreign bodies and carry them to the kidney and spleen. (Unfortunately, many bacteria have the ability to overcome the inflammation response, either by releasing toxins that destroy the ingesting white cells or by producing an agent that dissolves the fibrin walls and allows them to spread throughout the body).

In the kidney and spleen, special proteins known as antibodies are formed that will act specifically against each particular 'intruder', more precisely known as an 'antigen'. The antibody production process may take up to two weeks or more. Each antibody attaches to its specific antigen and acts against it in a number of ways: it may detoxify the antigen so it can be ingested by the white blood cells; inactivate its reproductive system so that it cannot proliferate; or trigger a series of blood components (know as the complement pathway) that help to destroy the whole antigen cell.

If the specific antigen has been previously encountered, the fish's immune system will react much more quickly; specific antibodies will already exist and will multiply extremely rapidly upon contact with that particular antigen. This is the principle behind vaccination, in which introducing a detoxified disease organism enables the fish to build up specific antibodies and thus increase its chances of survival during a bout of a specific disease.

The effectiveness of a fish's immune system is related to its environment. In low temperatures, the immune reaction is slowed down and, if the pathogens are not similarly hampered, then death may result. Fish tend to exhibit behavioural 'fever symptoms', showing preference for warmer water when under a disease attack. Environmental pollution reduces a fish's immune response.

CHAPTER 3

UNDERSTANDING WATER CHEMISTRY

As we have seen in Chapter 2, fish are basically fluids living within a fluid medium separated only by a fairly permeable membrane. It is impossible to underestimate how close the relationship is between fish and their environment. What this means is that any change in the properties or state of the water around a fish has an immediate and profound impact on its physiology. In fact; fish are much more vulnerable to the changing state of their environment than are terrestrial animals. However, two factors offset this apparent vulnerability. Clearly, having evolved over millions of years, fish have become superbly adapted to survive in their unique environment and can, to some extent, compensate and adjust to changing conditions. Also, the natural aquatic environment is generally quite stable. Water has a number of properties that, in general, ensure that changes occur very slowly, thus allowing fish time to adjust their physiological functions.

The nature of water

Pure water is simply two atoms of hydrogen and one of oxygen (H_2O). However, it acquires many other characteristics due to its remarkable properties as a solvent. Water is influenced by the nature of the atmosphere through which it falls as rain and the chemical nature of the soil, earth and rocks through and over which it passes on its way from high ground down to the sea. Evaporation, dilution by rain, biological factors and man's activities all contribute further to the final characteristics of a body of water.

The interaction of all these variables means that there is a tremendous variety of aquatic environments in the world, each with radically different physical and chemical characteristics. Fish and other aquatic creatures have evolved in all these different aquatic environments, adapting their physiology so that they are uniquely suited to live in water with specific characteristics.

The physical and chemical characteristics of water which have the most profound influence on aquatic life are its pH (acidity or alkalinity), hardness and salinity (both a function of the amount and type of dissolved salts), temperature, dissolved carbon dioxide and oxygen content, and the dissolved content of toxic materials, including organic matter such as nitrite, plus heavy metals and synthetic chemicals.

Left: Even as the water in this mountain stream begins its journey to the sea, its chemistry is being influenced by the rocks and soil over which it flows, thus dictating the living organisms that can thrive within it.

Understanding water chemistry and its effect on fishes is essential for good fishkeeping because it helps you, the aquarist, to provide the correct water conditions for the fishes in your care and to keep these conditions constantly perfect. If you take time and care in this aspect of fishkeeping you will be rewarded with less problems in terms of fish health. As in all things, careful prevention is far better than cure.

In this chapter, we explore all aspects of water chemistry, reflecting how each of the major characteristics of water influence fish physiology and health, including how fish have managed to cope with different extremes around the world.

Below: The natural influences on the pH value of water are many and varied. Even as rain falls through the air, it begins to lose its neutrality. It is further modified by rocks, soil, organic debris, pollution and the metabolic processes of photosynthesis and respiration.

Natural influences on pH value

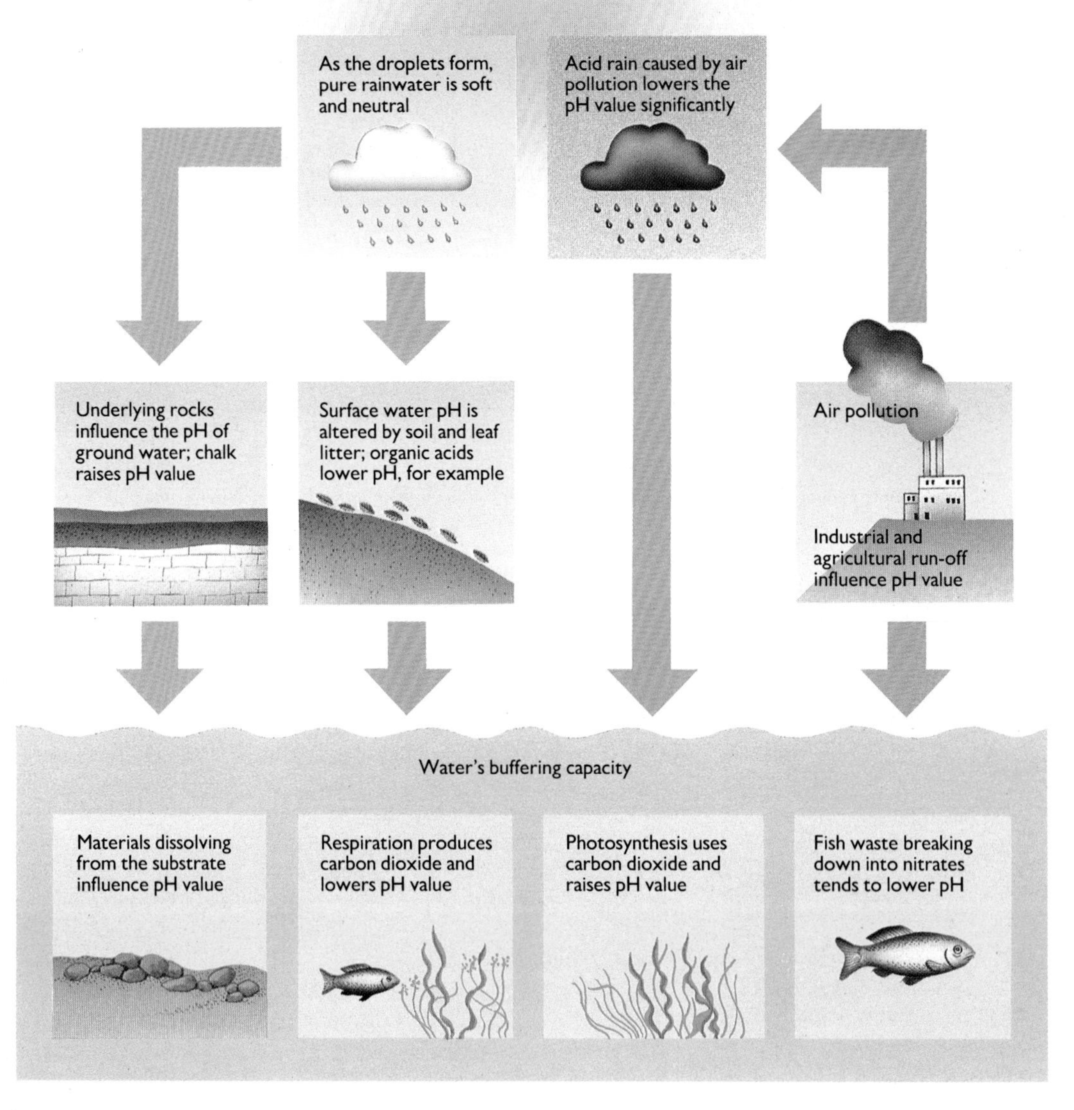

The pH scale

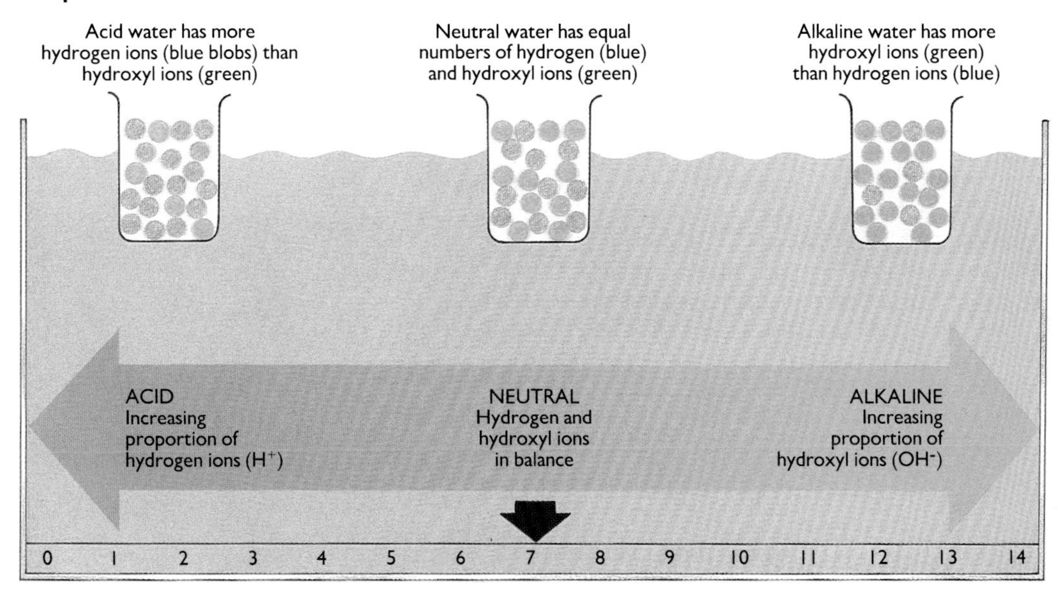

Above: The degree of acidity or alkalinity of water relates to the ratio of hydrogen ions to hydroxyl ions.

Below: Such a broad meandering stream rich in plant and animal life is open to many influences on its pH value, not least of which is agricultural run-off.

The pH value of water

In simple terms, pH value is a way of indicating whether a mass of water is acid, alkaline or neutral. To understand what this means in chemical terms, we need to refine our 'model' of water as two hydrogen atoms plus one of oxygen. Water actually 'exists' as free positively charged hydrogen ions (H^+) and negatively charged hydroxyl ions (OH^-) in varying proportions. The pH value of a sample of water is a measure of the proportion of hydrogen ions to hydroxyl ions. If these ions are present in equal numbers, the water is said to be neutral and is given a pH value of 7. As hydrogen ions progressively exceed the number of hydroxyl ions, the water becomes more acidic and the pH value falls below 7. If, however, hydroxyl ions progressively exceed the number of hydrogen ions, the water becomes progressively more alkaline and the pH value rises above 7 to a maximum of 14. The pH scale is logarithmic, which means that a one-unit shift registers a ten-fold change in hydrogen ion concentration. This means that a comparatively small rise or fall in pH value actually reflects a fairly substantial change. Note: pH is usually cited to one decimal place – e.g. pH 7.2, etc.

Many compounds added to water tend to break up (ionize) into their constituent ions and contribute either hydrogen or hydroxyl ions, therefore changing the pH value. Natural acidification of water occurs through the leaching of mineral acids, organic acids and mineral salts. Carbon dioxide gas from the air also has an acidifying influence because it readily associates with water to form carbonic acid. Therefore, as we shall see later, processes such as respiration, photosynthesis, turbulence and aeration, all of which alter carbon dioxide levels, also affect the water's pH.

Pollution can have a profound effect on pH values. In the wild, industrial waste containing free acids and certain metal ions lowers the pH value, while sewage and agricultural run-off can indirectly cause the pH value to rise by stimulating vast algal blooms that remove carbon dioxide through photosynthesis.

In fishkeeping, probably the most important factors that alter pH value are biological filtration, respiration and plant metabolism. Let us consider each of these processes in turn.

Biological filtration consists essentially of two processes, both carried out by colonies of bacteria in the filter medium:

- Nitrification, in which toxic substances, such as ammonia (NH_3), in fish wastes are broken down into nitrites (NO_2-) and then into less harmful nitrates (NO_3^-);
- Denitrification, which occurs in certain special filters, whereby nitrates are converted, eventually, to free nitrogen gas and oxygen.

These processes exert opposing effects on the pH value of water. The hydrogen and nitrate ions released during nitrification in effect produce nitric acid (HNO_3) and thus lower the pH value. Denitrification, by removing the acidifying nitrate ions, causes the pH value to rise. (See pages 57-62 for more details on these compounds and their importance in water chemistry.)

Respiration is the gaseous exchange at the heart of metabolism and has the effect of lowering the pH value (i.e. acidifying the water) because carbon dioxide (and therefore carbonic acid) is produced as a waste product. Aerating the water and introducing turbulence drives off the carbon dioxide and reduces this effect.

Plant metabolism uses up carbon dioxide during the process of photosynthesis and nitrate for growth, both of which have the effect of raising the pH value. If no free carbon dioxide is available for photosynthesis, the carbon dioxide is extracted from the bicarbonate ions (HCO_3^-) in the water, which reduces the water's buffering capacity. This is why it is vital to avoid excessive plant and algal growth. Summer algal blooms in ponds, for example, can severely disrupt the pH value of the water.

As a stabilizing influence on all these potential disruptions to the pH value, water generally contains substances called buffers that suppress fluctuations of hydrogen ion concentration and thus keep the pH value reasonably steady. The buffering capacity of water is based on the total content of substances that will counteract the pH value-raising effects of positive hydrogen – in effect, on its alkalinity. (Buffering capacity is closely related to water hardness, as we shall see later.) Water's alkalinity is due to the presence of hydroxides, carbonates (CO_3^{--}), bicarbonates (HCO_3^-) and, to a lesser extent, borates, phosphates, arsenates, silicates and ammonia. A high alkalinity results in a high pH value. Sea water is generally better buffered than most fresh waters due to its high salt content and, not surprisingly, it has a pH value in the region of 7.9-8.3.

Fish and pH value

Fish have evolved to exploit aquatic environments of widely differing pH values, from pH 5 to pH 9.5. The majority of freshwater fish species live in environments with a pH level between 6 and 8. In general, environmental pH levels remain fairly constant, with a maximum daily range of only a few fractions of a pH unit. Those fishes which have evolved to deal with acid environments, i.e. with a pH value below 7, are called acidophiles;

Above: These brilliant rasboras (*Rasbora einthoveni*) from Southeast Asia thrive in slightly acid water i.e. with a pH range of 6-6.5. Such fish are acidophiles.

Below: By contrast, these cichlids (*Haplochromis moori*) are known as alkalophiles because they thrive in the alkaline water of Lake Malawi, at pH 8-8.5.

rasboras and discus are examples of this type of fish. Fish species which have evolved to cope with alkaline conditions, i.e. with a pH value above 7, are called alkalophiles; many barbs and the African Rift Lake cichlids are alkalophiles.

The major physiological relevance of pH to fish is their need to maintain a constant internal pH level and an acid/base (i.e. acid/alkali) balance in the blood. Fish counteract pH changes by using basic bicarbonate ions and/or acidic carbon dioxide. If the blood becomes too acidic due to an excess of carbon dioxide, either in the environment or as a result of metabolic action, the blood plasma level of bicarbonate ions rises to buffer the pH level back up to its normal value. Conversely, the addition of carbon dioxide or the removal of bicarbonate ions helps to lower the blood pH level. These reactions are speeded up by a hormone called carbonic anhydrase in the blood and gill membranes.

Fish species vary in their ability to deal with pH changes and adults tend to cope better with pH changes than do fry or eggs. Obviously, the processes for maintaining a steady blood pH level and acid/base balance are more highly developed in fish which live in environments towards the two extremes of pH value and in fish which are known to be less sensitive to pH changes. If the pH value of water exceeds a fish's normal narrow range for a long period of time or radical changes in pH value occur within its normal range, then the fish shows symptoms of either acidosis or alkalosis.

Acidosis generally occurs below pH 5.5, although this depends on the species of fish and the pH value of its natural environment. The behavioural reaction of fish to such acidic conditions varies according to whether the pH change is rapid and acute or slow and chronic. In the former case, fish become highly excitable, making rapid swimming movements, gasping and tending to jump; death follows fairly rapidly. In the latter case, the effects are more subtle, resulting in slow death with few obvious behavioural symptoms. The gasping results both from the reduced oxygen-carrying capacity of haemoglobin in acidic conditions and the excess mucus produced by the irritated gills. Below pH6, colloidal iron (i.e. iron associated with complex organic molecules) is deposited on the gill, leaving dark grey deposits and further inhibiting gaseous exchange.

Acidity irritates not only the gills but also the skin and all external body surfaces, resulting in excess mucus production, (the skin showing a milky turbidity), and reddened skin areas, especially in the ventral region. Low pH levels also increase a fish's susceptibility to disease, particularly to bacterial infections.

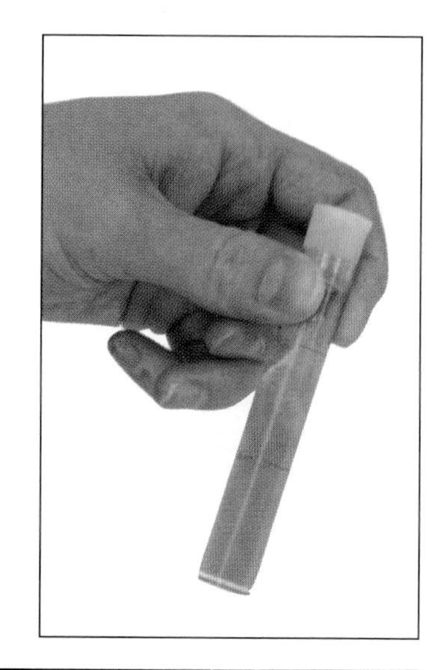

Alkalosis generally occurs above pH8 to pH9, depending on the species of fish; acidophiles, for example, experience alkalosis at pH8, while marine fishes do not suffer any symptoms until the pH level exceeds 9. The behavioural and physiological symptoms involved are much the same as for acidosis. In addition, the fish's

pH	Aquatic environment
6.0	Blackwater Amazon tributary
6.3	Malayan river
7.0	Lake Chad (middle)
7.4	Mexican river
8.3	Florida coral reef (summer)
8.5–9.0	Lake Tanganyika

pH and hardness ranges of some common fish	
Acidophiles pH 6-7.5 total hardness up to 50mg/litre $CaCO_3$ Angelfish (*Pterophyllum scalare*) Clown loach (*Botia macracantha*) Discus (*Symphysodon discus*) Harlequin (*Rasbora heteromorpha*) Killifishes (*Aphyosemion* and *Epiplatys*) Neon tetra (*Paracheirodon innesi*) Ram (*Papillochromis ramirezi*) Red piranha (*Serrasalmus nattereri*) **Medium range pH 7.5-8** total hardness 150-300mg/litre $CaCO_3$ Black Molly (*Poecilia* hybrid) Common carp and koi (*Cyprinus carpio*) Goldfish (*Carassius auratus*) Guppy (*Poecilia reticulata*) Platy (*Xiphophorus maculatus*) Rainbowfish (*Melanotaenia* and *Bedotia*) Swordtail (*Xiphophorus helleri*) **Alkalophiles pH 8-9** total hardness 300-450mg/litre $CaCO_3$ African Rift Valley Lake cichlids (Such as *Haplochromis, Hemichromis, Julidochromis, Lamprologus, Melanochromis, Pseudotropheus, Tropheus,* etc.) Marine fish and invertebrates	**Tolerant species pH 6.5-8.5** total hardness 50-350 mg/litre Blue acara (*Aequidens pulcher*) Corydoras catfishes (*Corydoras*) Firemouth cichlid (*Thorichthys meeki*) Gouramis (*Colisa* and *Trichogaster*) Red-tailed black shark (*Labeo bicolor*) Rosy barb (*Barbus conchonius*) Siamese fighting fish (*Betta splendens*) Tiger barb (*Barbus tetrazona*)

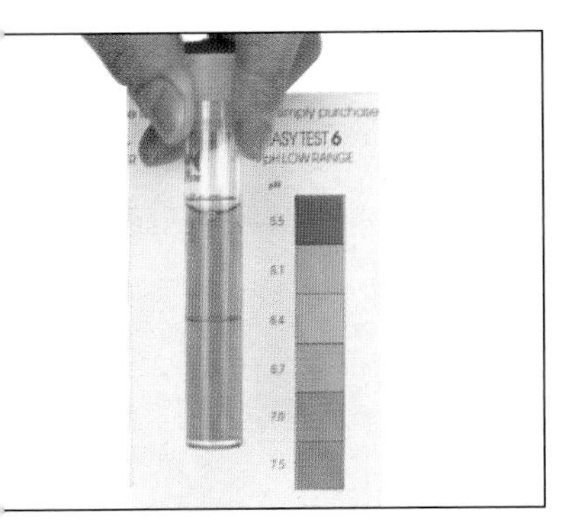

Left and above: Using this pH test kit involves adding a reagent to a water sample and comparing the colour change to a printed chart.

Left: Many fishes thrive in narrow ranges of pH and hardness values; others are more tolerant.

Below: An easier, although initially more expensive, way of recording the pH value of water is to use an electronic pH meter.

gill and fin tissues are destroyed. A further problem experienced at high pH levels is increased toxicity of ammonia, a factor that we will examine in more detail on page 57.

Controlling the pH value of water

It is vital to provide the correct pH range for individual fish species, if possible matching the natural range to which they have become physiologically adapted over millions of years. However, the more important point is to maintain the water's pH value at a relatively constant level. You can monitor pH levels very simply and relatively accurately with the readily available liquid, paper and tablet kits or by using special electronic pH meters. If you need to make any intentional alteration to the pH level then make these gradually in steps of not more than 0.3pH units per 24 hour period. This gradual pH change allows the fish to adapt physiologically.

Providing a stable pH level for fishes that thrive in neutral to alkaline conditions is made relatively easy by having plenty of alkaline buffering salts in the water. Providing stable pH conditions in soft acid water requires greater care. In this case, useful precautions against a variation in pH level include providing adequate aeration and turbulence to ensure that much of the carbon dioxide produced in respiration does not lower the pH level still further. (On the other hand, avoid excessive plant growth because this will raise the pH value by removing carbon dioxide and nitrate, both acid-forming substances.) The acid-forming activities of biological filtration, i.e. the process of nitrification that produces nitrates and then nitric acid, should be countered either by using denitrification, which breaks nitrate down into nitrogen and oxygen, or by making regular partial water changes to maintain good water quality.

Creating acidic conditions

There are a number of methods of producing acid water for acidophile species of fish. The most common method is to use aquarium peat. Peat varies in its acidifying properties, and soft water low in buffers acidifies more easily than relatively hard water. Suitable peat should be available in aquarist shops; otherwise, experiment with peat from garden centres, but do ensure that it does not contain any chemicals. Place the peat in a loose mesh bag, allowing one handful of peat per 5 litres (approximately a gallon), and treat the water in a separate tank for about two weeks before adding it to the main aquarium. There are now also some low pH buffering salts available. These are relatively easy to use and help to stabilize the pH level but they have the slight disadvantage of increasing the dissolved salt content of the water. Since most acidophiles are also softwater fish, and thus used to very low levels of dissolved salts, this effect can be undesirable.

It is important to remember that acid water has the effect of radically reducing the efficiency of biological filtration. This is because the optimum working pH level for nitrifying bacteria is around pH7; at pH levels below this their action is suppressed, especially below pH6. Thus, if a pH value below 6 is absolutely necessary, consider using chemical filtration media, such as zeolite (an ammonia-removing substance) and making more frequent partial water changes.

Ensure that all decorative and substrate materials used in low pH aquariums are as inert and insoluble as possible. This is because soluble materials, such as calciferous rocks, will dissolve and release bicarbonate and carbonate salts that will increase the alkalinity of the water and hence raise the pH value.

Creating alkaline conditions
When alkaline water is required for keeping alkalophiles, use calciferous material, such as calcium-rich filter media, tufa rock, coral sand and crushed shells, for decoration and as a substrate. These calciferous materials slowly dissolve, buffering any acidification processes occurring in the aquarium. After a period of time, however, they gradually lose their buffering effect because they become coated with organic and mineral deposits.

Commercially available chemical buffering additives can be used to ensure that the water's pH value stays high and that its alkali reserve, which counters pH changes, is not exhausted. These buffering chemicals, usually based on sodium bicarbonate, ensure that a major imbalance of ions does not occur, especially in seawater situations. Maintaining a stable high pH environment for Rift Valley cichlids and all marine life is particularly crucial. This is because these creatures come from very large bodies of water with massive innate buffering capacities and consequently have not developed the ability to adapt to changing pH levels.

Water temperature and its effects on life processes
Water has a high specific heat capacity, which means that temperature changes are resisted and occur relatively slowly. In terrestrial environments, daily temperature ranges of 15°C(27°F) due to solar heating are not unusual. Most aquatic environments, however, will vary only by 3–4°C (5–7°F). Seasonal temperature changes in aquatic environments occur slowly over a matter of months. The only extreme temperature changes occurring in most water bodies result from the influx of large quantities of cold rainwater or melting ice, or even thermal effluent from factories or power stations. Therefore, fish have evolved in an environment where the temperature remains relatively stable and any changes that occur do so slowly.

Water density varies with temperature in the same general way as other materials, i.e. cold water is denser than warm water and has a tendency to sink. Therefore, the warmest water is usually found at the surface and water temperature falls with increasing depth. However, water reaches it highest density at 4°C(39°F), and below this temperature it becomes progressively less dense, until at 0°C(32°F) it freezes and the resulting ice floats on the surface. This remarkable – and unique – characteristic allows aquatic creatures to survive in cold climates because water freezes from the surface downwards and leaves a layer of water at 4°C at the bottom of the water body. Even if the temperatures are not low enough for ice formation but nevertheless in the region of or below 4°C, the lower water layers provide a natural 'slightly warmer' refuge for relatively inactive fishes to spend the winter months.

Fish are ectothermic, which means that they have little ability to

Above: The rock goby (*Gobius paganellus*) can cope with fluctuating temperature changes, a physiological adaptation to living in shallow coastal rock pools.

Left: Creatures that thrive in rock pools such as these must withstand quite severe variations in their environment, particularly in temperature and salinity.

Below: Like many marine fishes, the superb flame angelfish (*Centropyge loriculus*) from the Pacific Ocean, is intolerant of environmental changes.

physiologically maintain a constant body temperature; in most fish it is usually the same as the temperature of the surrounding water. Even so, fish are found throughout the aquatic habitats of the world, having acclimatized and evolved to survive over a tremendous range of temperatures. Species are found beneath the polar ice caps, where they withstand subzero temperatures thanks to a natural 'antifreeze' glycoprotein in their blood. At the other extreme, some tilapia live in the hot springs of the East African Rift Valley, where the highly mineralized water helps them to survive temperatures of around 38°C (100°F). (As we explain on page 42, the process of osmoregulation becomes disturbed at high temperatures, and highly mineralized water helps to counteract this effect by reducing osmotic stress.) In general, fish species are loosely divided in terms of temperature range into warmwater species from the tropics that live at temperatures over 24°C(75°F) and coldwater species from high latitudes, where seasonal averages are well below that level.

In any one environment, fish acclimatize to a relatively narrow temperature range. If the temperature moves outside this range for a continuous period or changes rapidly within the accepted range then it causes stress to the fish. Although this is true as a general statement, fish species do vary in their ability to withstand temperature change. The so-called 'eurythermic' species, such as the rockpool gobies (*Gobius sp.*), can adjust to temperature changes better than can 'stenothermic' fishes, such as the marine angelfish (Pomacanthidae).

We shall be looking at the physiological implications of temperature change in more detail later on, but in general the major effects on fish are: an alteration in metabolic rate (a 10°C/18°F rise in temperature doubles the metabolic rate, for example), a disturbance of respiration (warm water holds less oxygen than cold water); a blood pH imbalance; and a breakdown in osmoregulation function. Sudden temperature changes also often cause swimbladder problems. Temperature also directly affects growth and development, so unsuitable temperatures can lead to fish growth being stunted, larvae failing to feed or eggs hatching prematurely. In general, adult fish are better able to deal with sudden changes in temperature than are their larvae or eggs. (This parallels the relative response of adults and young to pH levels, as explained on page 37.)

Fish attempt to cope with sudden thermal stress either by moving to a new location with a more suitable temperature or, failing this, attempting to acclimate or compensate physiologically. This process of adjustment generally follows the same pattern. During an initial shock stage, the metabolic rate overshoots (i.e. becomes excessively above or below normal) and then stabilizes over a period of hours to a new steady state. The fish's physiological functions then adjust slowly over a period of days, or even weeks, depending upon the degree of temperature change, to suit the new environmental temperature. The degree and rate of adjustment depends upon the species of fish, its sex, its state of nutrition, the salinity of the water and the dissolved oxygen levels. If the fish is stenothermal or the temperature change is excessive, the effects can prove directly lethal or the resultant stress can cause reduced resistance to disease. Thus, sick fish are more susceptible to the effects of marked temperature change than are healthy fish.

The physiological effects of temperature increase and decrease are slightly different; in general, fish adjust better to an upward rather than a downward temperature change of the same magnitude. Here, we look briefly at the effects of increasing and decreasing temperature.

Increasing temperature results firstly in a vicious spiral of increasing metabolic rate, and therefore a growing demand for oxygen, in the face of a falling oxygen content of the water (since rising temperature reduces the quantity of dissolved oxygen that water can hold in solution). The oxygen demand for an American perch (*Perca flavescens*), for example, increases tenfold between 5 and 25°C(41-77°F). Physiologically, the oxygen deficit resulting from high temperatures causes an increased production of adrenalin and a faster heart rate. The problem is exacerbated by a parallel reduction in the blood's oxygen transporting capacity. Behavioural changes noted in fish during high temperature thermal stress include heightened activity, a loss of equilibrium and increasing ventilation rates.

High temperatures cause denaturing of body proteins and enzymes, and cells damaged by this process produce toxic metabolites. High temperatures also raise the inherent toxicity of certain substances, most notably heavy metals and ammonia. And osmoregulatory problems occur at high temperatures because lipids change state in cell membranes, causing an increase in the permeability of the cells. This is especially crucial in the gills.

Under extreme thermal stress, a fish slips into a coma as the central nervous system shuts down. The maximum temperature a fish can survive depends on the species, the temperature to which it is normally acclimatized, the amount of dissolved oxygen in the water and the level of toxins present.

Falling temperature also causes metabolic problems; below 15°C(59°F), metabolism is temperature limited and it is possible that energy liberation may fall below the level required to keep the body functioning if the fish is not physiologically adapted to low temperatures. (Above 15°C, metabolism is limited primarily by the water's dissolved oxygen content.) Low temperatures may also result in insufficient uptake of oxygen, a condition known as hypoxia, in spite of the fact that cold water holds more oxygen than warm water. Oxygen uptake is hampered by a falling respiration rate and a slowing of the heart rate. Sudden cooling can also result in degeneration of red blood cells and the subsequent loss of oxygen-carrying haemoglobin, with obvious detrimental effects on respiration efficiency.

Low temperatures cause chronic osmoregulatory problems; the gill membranes become permeable, the salt pump in the gills stops working and kidney failure occurs. Low temperature thermal stress behaviour usually includes loss of equilibrium and sudden violent spasms. In extreme cold shock, fish generally slip into a coma when the central nervous system ceases to function. This can be a reversible condition, if circumstances improve radically.

Below: This simple bar graph shows how increasing water temperature markedly accelerates oxygen consumption, in this case in an active goldfish.

Oxygen consumption of an active goldfish at different temperatures

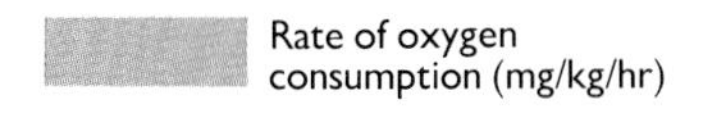

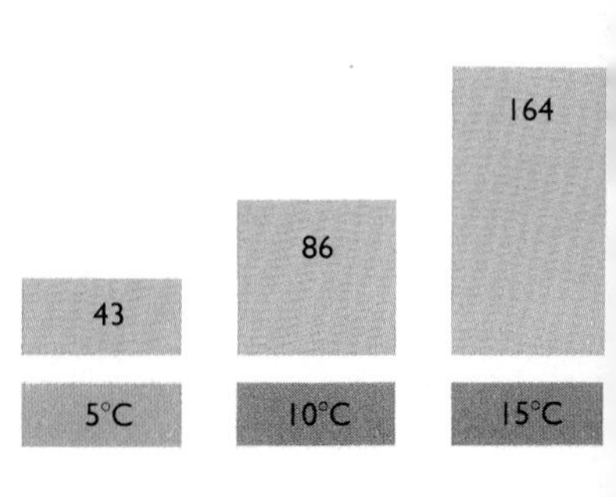

Above: The warm, soft, acidic waters of this Malaysian stream are low in oxygen. Many of the native fish have accessory respiratory organs to use atmospheric air.

Unsuitable temperature or radical temperature changes usually result in the reduced effectiveness of the immune system. For instance, antibody production takes longer at lower temperatures. An interesting behavioural response to disease is the attempt fish make to find warmer water when under disease attack, which is a kind of behavioural fever response. Many fish diseases have definite temperature associations, not only in terms of the direct effect of temperature on the fish and its immune system but also on the virulence of the pathogens. For example, a disease episode caused by *Aeromonas* bacteria resulted in 14 percent mortalities in a fish population at a water temperature of 4°C(39°F), while at a temperature of 21°C(70°F) the mortalities reached 100 percent due to increased bacterial virulence.

Providing the correct temperatures

As we have seen, it is important to maintain the temperature of the water within the natural range of the species being kept and as stable as possible. Using modern heater-thermostats, and even cooling systems, plus easily readable thermometers make this task so much easier. Heating requirements for aquariums vary according to the volume of water they hold and the ambient temperature of the room in which they are kept. Cooling systems specifically for aquarium use are also available. Alternatively, home-made devices can be quite effective and usually involve passing water from power filters through a beer chiller or coils of tubing placed in a refrigerator. Trial and error is the only way of achieving a constant temperature. If thermal stress is unavoidable when dealing with freshwater fish, adding a physiological salt – i.e. one that contains the main ions essential for bodily functions, such as calcium (Ca^{++}), chloride (Cl^-), potassium (K^+) and sodium (Na^+) – up to a concentration of 0.2-0.5 percent will reduce the detrimental effects on the fish. This is because the salt reduces the load on the osmoregulatory system, which is one of the major functions affected by thermal stress. At high temperature, ensure that the water has adequate aeration to maximize its oxygen content.

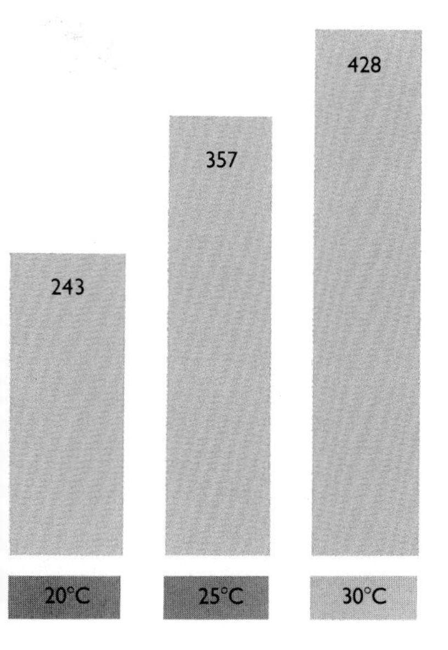

Dissolved salt content

Because of water's remarkable solvent qualities, natural water contains a wide variety of dissolved substances. The type and amount of dissolved salts found in fresh water generally depends on the water's source and on the chemistry of the rocks and soil over and through which the water has passed before reaching the watercourse. Over 95 percent of all dissolved substances in natural water consist of eight ions, four negatively charged ions: chlorides (Cl^-), sulphates (SO_4^{--}), carbonates (CO_3^{--}) and bicarbonates (HCO_3^-), and four positively charged ions: calcium (Ca^{++}), magnesium (Mg^{++}), sodium (Na^+) and potassium (K^+). The remaining dissolved substances are present at trace levels. These include ions such as phosphate, nitrate and silicate, and minerals such as iodine, copper and zinc, as well as other metal ions. A careful balance of these dissolved substances is essential for aquatic life to exist. The relative concentration of these ions determines two important characteristics of water: hardness and salinity. Here, we examine each of these parameters in more detail.

Water hardness

Water hardness is generally referred to in the analysis of freshwater aquatic environments and is concerned only with part of the total dissolved salt content. It is a measure of the quantity of certain metallic ions present in the water, notably calcium and magnesium. Of these, calcium is the most significant, outnumbering magnesium ions by between three and ten times. To a much lesser extent, barium, strontium, iron, copper, zinc and other metallic ions also have an influence on water hardness. These ions are generally present in three main forms: hydroxides, carbonates and bicarbonates, while smaller quantities of sulphates, chlorides, silicates, phosphates and borates also exist.

The total water hardness value (also known as general hardness and designated GH) refers to the total content of all these combinations of salts and can be subdivided into temporary hardness, which is removed by precipitation if the water is boiled, and permanent hardness, which remains in solution after boiling. Temporary hardness (also known as carbonate hardness and designated KH) usually forms the major portion of the total hardness. Certain calcium and magnesium salts contribute to alkalinity as well as water hardness. Alkalinity and water hardness have quite a complex relationship, but in simple terms alkalinity closely reflects temporary hardness. This is because temporary hardness is caused mainly by bicarbonate ions, which are also chiefly responsible for the water's alkalinity. The permanent hardness portion of total hardness consists mainly of carbonate, chloride and sulphate salts.

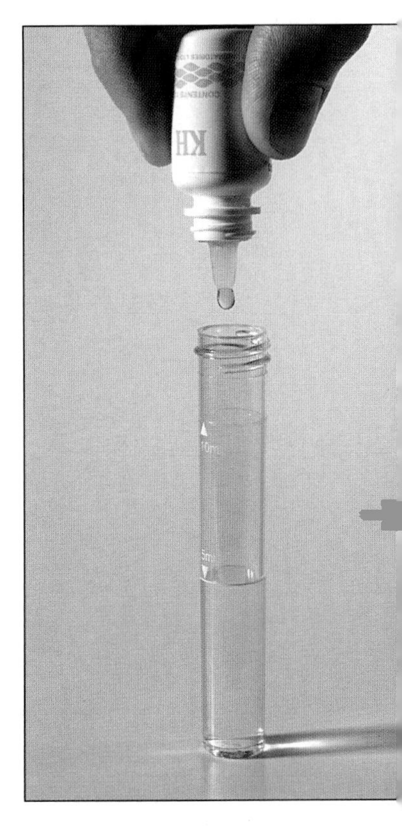

Above: Testing for carbonate hardness (KH) using this kit involves adding drops of a reagent to a sample of water. The first drop turns the sample blue.

Measuring water hardness

Water hardness is measured in a number of different ways. The most common methods are chemical titration techniques that involve adding two chemicals to a sample of the water until a defined colour change occurs. Different titrations are used to calculate total and temporary/alkaline hardness. Total hardness can also be determined using a soaping reagent; this is based on the fact that soap foams much more easily in soft water than in hard, and therefore more soap has to be added to a hardwater sample before a permanent foam forms.

Measuring the water's electrical conductivity also gives an indication of its hardness. This is because the higher the ion content the greater is its capacity to pass an electric current. Using an electronic test meter to register water hardness simply involves

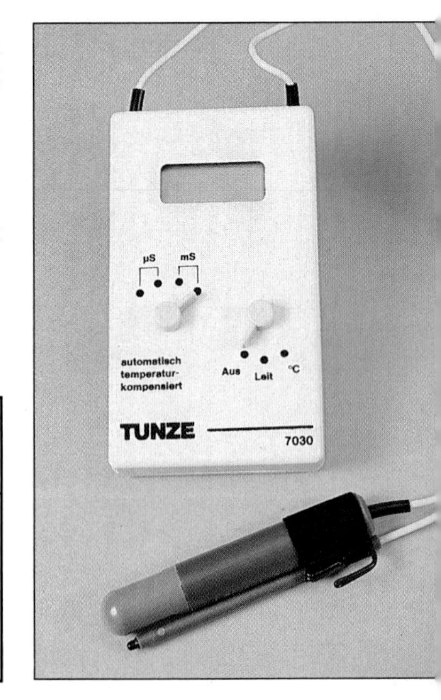

Comparing water hardness scales

Scale	Origin	Equivalent in terms of mg/litre $CaCO_3$	Conversion factor to mg/litre $CaCO_3$
° hardness	USA	1mg/litre $CaCO_3$	–
° Clark	UK	14.3mg/litre $CaCO_3$	14.3
° dH	Germany	17.9mg/litre $CaCO_3$	17.9
° fH	France	20mg/litre $CaCO_3$	20.0

Above: Continue to add drops one at a time and shake the sample. The number of drops needed to change the colour to yellow is equivalent to hardness in °dH.

Left: Using a conductivity meter provides an instant readout of total water hardness. In general terms, hard water conducts an electric current more readily than soft water. The test simply involves dipping the probe into the water and converting the readout in micro Siemens into a hardness scale.

Water hardness in comparative terms

Mg/litre $CaCO_3$	°dH	Considered as
0-50	3	Soft
50-100	3-6	Moderately soft
100-200	6-12	Slightly hard
200-300	12-18	Moderately hard
300-450	18-25	Hard
Over 450	Over 25	Very hard

dipping the probe into the water sample and noting the conductivity reading, which is expressed in units called micro Siemens (μS). This method, although quick, needs relatively expensive equipment and cannot differentiate between the permanent and temporary hardness components of the sample. Water hardness is expressed in a confusing array of scales, although a standardizing influence is to express them in terms of milligrams per litre of calcium carbonate (mg/litre $CaCO_3$), which is also equivalent to parts per million of calcium carbonate (ppm $CaCO_3$). The scales and their conversion factors are shown in the table at the bottom of page 44.

How water hardness affects fish

Water hardness affects freshwater fishes in terms of osmoregulation. Since hard water has a higher concentration of salts than soft water, for example, the osmotic difference between the fish's internal environment and the surrounding water is smaller. In hard water, therefore, the osmoregulatory system has a reduced work load replacing ions lost from the blood. (The presence of calcium in the water also decreases cell permeability, which reduces ion losses and water influx.) Conversely, fish in soft water need to have more efficient osmoregulatory systems and expend more effort to maintain their internal salt/water balance.

Water hardness also affects the regulation of blood calcium level, which also depends on diet. Hardwater fish cope with excess blood calcium by using an efficient system for excreting calcium, governed by a hormone called calcitonin. Softwater fish need to obtain more calcium from their diet and also use bones as calcium reservoirs to ensure that blood calcium levels remain constant.

Freshwater fishes have adapted to thrive in an immense range of water hardness values, from the soft acid waters of the Amazon River, with less than 50mg/litre calcium carbonate (3°dH), to the hard alkaline waters of the African Rift Valley lakes, with hardness values exceeding 330mg/litre calcium carbonate (18.5°dH). Within any particular environment, however, fish have tuned their physiological functions to cope efficiently with a fairly narrow range of water hardness levels and therefore osmotic pressure. Altering the hardness values outside this range or disrupting the major ion composition of water hardness will lead to extreme osmotic stress and other physiological malfunctions. As an example, if bicarbonate ions exceed normal levels then bicarbonate excretion by the gills fails, leading to an alkaline pH shift of the blood and symptoms of alkalosis (see page 38). High bicarbonate levels also alter kidney functions, resulting in calciferous deposits in the kidney tubules. Fish eggs are also affected if they are put in abnormally hard water; they fail to 'harden'. This may sound contradictory, but the egg-hardening process involves the uptake of water from the surrounding medium so that the egg becomes turgid (i.e. 'hard'); a higher than usual hardness level reduces the osmotic gradient between the inside and outside of the egg and thus decreases the inflow of water.

Different fish species have a varying resistance to changes in water hardness, depending on their ability to alter their osmoregulatory process to changes in osmotic demand. Most fish can be acclimated slowly to abnormal water hardness. In general, however, they will be under unnatural stress and will not achieve optimum performance in growth, breeding or disease resistance.

Water hardness also has a distinct effect on the toxicity of certain minerals and pesticides. In general, harder water or high calcium levels reduce the toxicity of these substances. However, as we shall see on page 57, ammonia is more toxic in hard water because of its alkalinity (which also means that it generally has a high pH value, as explained on page 36).

Above: Water quality can have noticeable effects on fish health and appearance. These discus, for example, are clearly paler in unsuitably hard water.

Controlling water hardness

It is clear that it pays to provide your fish with water of the same hardness as they have evolved to deal with in their natural environment. Preferred water hardness ranges for individual fish species are easily available in standard reference books. Use this information to make sensible choices for your aquarium. If you aim to set up a community aquarium, be sure to select fishes and other

Left: A healthy discus in ideal soft, slightly acid water conditions. Although coloration in discus varies with species and age, it is clear that the best water conditions promote radiant health and vivid markings.

creatures that share similar water chemistry requirements, most importantly the same range of hardness and pH levels – which are fairly closely linked anyway. Mixed community aquariums with softwater-loving tetras and rams living alongside hardwater species such as guppies and swordtails can only be a compromise situation. It would be storing up disease trouble for one or both groups of fish without deriving the best from either.

To suit certain groups of fishes, it is often necessary to change the hardness value of your local tapwater. Here, we look at simple ways of providing hard and then soft water suitable for fishkeeping.

Making water hard

Most tapwater is extracted from calciferous aquifers (i.e. from water that collects in underground formations in chalk-bearing or limestone rocks) and is thus fairly hard; where soft tapwater occurs it is a relatively easy matter to make it harder. This process generally involves adding calcium-based salts to the water until the desired hardness level is achieved. Check this by using a suitable test kit. Incorporating calciferous material in the aquarium substrate and decoration, as well as in the filter (where space allows), will keep the water hard.

Making water soft

Providing soft water is not as simple. Collecting rainwater – which is almost pure and, therefore, very soft – is one method. However, availability and a polluted atmosphere means that large storage facilities and careful filtration treatment are required to ensure satisfactory water quality. Rainwater also needs to be supplemented by some tapwater, since it is devoid of the trace elements essential for fishkeeping. However, this still results in a

Mixing water to the correct hardness

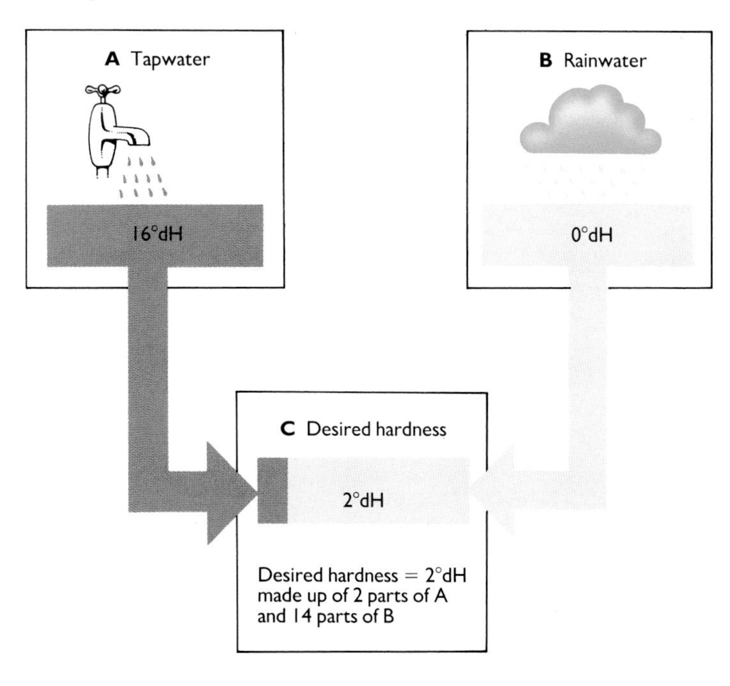

Right: Given two sources of water of known hardness, how do you mix them together to achieve the desired hardness for your fish? This diagram, a simplified version of the 'Pearson's Square', shows that the required proportions of tapwater (A) and rainwater (B) to create a specific hardness (C) are the difference between B and C of tapwater and the difference between A and C of rainwater. Thus, 2 parts of A at 16° dH plus 14 parts of B at 0°dH will produce a finished hardness of 2°dH. This works conveniently in the German °dH scale; to convert the result to mg/litre $CaCO_3$, multiply by 17.9.

poor balance of elements. Using the so-called 'Pearson's Square' (our simplified version is shown on page 47) will help in choosing the correct quantities of two sources of water with different hardnesses to produce the desired level of water hardness. Always take great care in providing consistent water quality.

A number of ion-exchange resins are available for fishkeepers to use and these do extract hardness from water. Normal deionising columns exchange calcium ions for sodium ions. However, this results in an excessive concentration of sodium ions in the treated water, which is unsuitable for most softwater fish. Two-column total demineralization columns replace positively charged ions, such as calcium (Ca^{++}) and magnesium (Mg^{++}), with hydrogen ions (H^+), and replace negatively charged ions, such as bicarbonate (HCO_3^-) and sulphate (SO_4^{--}), with hydroxyl ions (OH^-). The drawback with this technique is that sometimes toxic amines are produced and the soft 'pure' water again needs diluting with hard tapwater to provide the essential trace elements.

A partial deionizing system is available which removes only the bicarbonate hardness from the water. Since this makes up the majority of the total water hardness, the remaining hardness value rarely exceeds 50mg/litre calcium carbonate (equivalent to 3°DH). This method has the advantage of leaving most of the trace elements behind, making the addition of tapwater unnecessary and easing the task of providing consistent conditions. Partial deionization works by replacing the positively charged ions attached to the bicarbonate, such as calcium, with hydrogen ions. Although carbonic acid is produced in the process – and this acidifies the water – it readily breaks down and the carbon dioxide released is easily removed by vigorous aeration.

Reverse osmosis (RO) is another effective method of softening water. RO units can be purchased for home use, however they are fairly expensive and their output rates are slow. The process involves literally 'squeezing' pure water through a semi-permeable membrane by applying pressure in the opposite direction to the natural osmotic gradient. Peat is also a natural deionizer and has some water-softening properties.

Always use inert substrates and decorations in a softwater aquarium to ensure that they do not dissolve and raise the hardness value of the water.

Water salinity

Salinity is the measure of total dissolved material in a water sample. Water bodies can be differentiated into three major categories using salinity as a guide. These are: fresh water, which has relatively low salinity; sea water, which has a high salinity; and brackish water, which lies somewhere between the two extremes.

Salinity, which has the most relevance to marine fishkeeping, is generally measured in grams of dissolved material per kilogram (i.e. litre) of water, equivalent to parts per thousand. Sea water, for example, normally has a salinity of 35gm/litre (35 parts per thousand), although it varies with location around the world, and especially so in inshore waters. However, salinity is very complex to measure by direct methods, and its close relationship with the specific gravity of water is usually used as a guide instead.

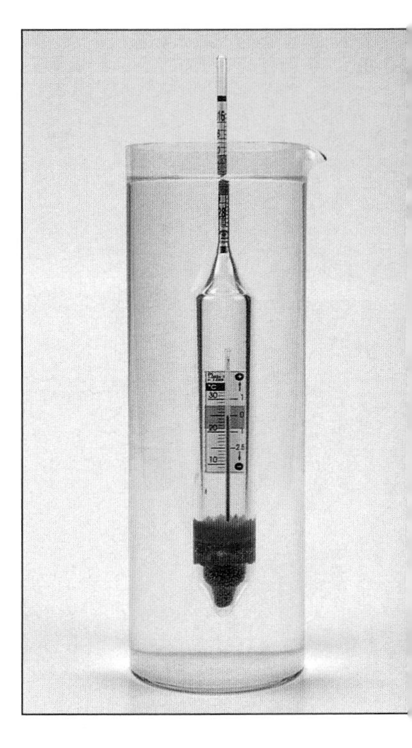

Above: Checking specific gravity with a floating hydrometer.

Specific gravity/salinity at 15°C(59°F)

	Specific gravity	Salinity (gm/litre)
	1.015	20.6
	1.016	22.0
	1.017	23.3
	1.018	24.6
	1.019	25.9
	1.020	27.2
	1.021	28.5
	1.022	29.8
	1.023	31.1
	1.024	32.4
	1.025	33.7
Sea water	1.026	35.0
	1.027	36.3
	1.028	37.6
	1.029	38.9
	1.030	40.2

Above: A swing-needle hydrometer in a marine tank provides a readout of the specific gravity of the water. Here, the needle indicates a reading of 1.020, which is lower than natural sea water but is ideal for keeping fishes in a marine aquarium.

Left: This table compares a range of specific gravity and salinity readings at 15°C(59°F). At this temperature, the typical sea water salinity of 35gm/litre is equivalent to a specific gravity of 1.026. At higher temperatures, the equivalent specific gravity falls; at 24°C(75°F), for example, 35gm/litre corresponds to a specific gravity reading of 1.024.

Specific gravity is the comparison of the weight of a water sample with the weight of an equal volume of distilled water at 4°C(39°F) – which is the point at which pure water reaches maximum density. Specific gravity is usually expressed as a ratio, with pure water being assigned a specific gravity of 1.000. Since adding dissolved substances to pure water not only raises its salinity but also increases its weight, specific gravity can be used as a direct measure of salinity. Sea water has a salinity of between 1.023 and 1.027, depending upon which sea it comes from. (The nominal salinity of 35gm/litre given above is equivalent to a specific gravity of 1.026 at 15°C/59°F and 1.024 at 24°C/75°F).

Specific gravity is measured with a hydrometer, traditionally a carefully weighted float that bobs higher in increasingly saline water, but now joined by swing-needle types that can be easier to read. Since specific gravity alters with temperature, most hydrometers are calibrated to be accurate only at a specific temperature – aquarium models normally at 24°C(75°F) – and correction factors are required if the temperature varies from this calibration level. Some fairly accurate modern hydrometers, however, claim to be relatively unaffected by temperature. Specific gravity can also be measured very accurately using an electrical conductivity meter (see page 44).

The effect of salinity on fish

The primary effect of salinity levels on fish physiology is directly related to their osmoregulation. (For full details of osmoregulation, see pages 26-27). Fish species have evolved to withstand a relatively narrow salinity range. Shifting the salinity outside these ranges results in excessive osmotic pressure, a physiological change in the body cells and a general stress response, which lowers their resistance to disease. Different species vary in their ability to survive salinity changes, depending on the degree to which they can adapt their osmoregulatory process. In general, fish found in brackish waters, such as scats (*Scatophagus argus*), and fish found in medium hard fresh water, such as blue acaras (*Aequidens pulcher*), survive salinity alteration better than those found in stable environments at the extremes of salinity, i.e. in soft fresh water or in full sea water.

Salinity in marine fishkeeping

The specific gravity of most of the oceans and seas of the world varies from 1.023 to 1.027, which at 15°C(59°F) is equivalent to a salinity range of 31.1–36.3gm/litre. In any one place, however, the salinity remains remarkably constant, due simply to the sheer volume of the seawater medium, which counters any processes that tend to alter the salinity. These include the concentrating effects of evaporation, which tends to increase salinity, and the influx of diluting fresh water such as rain, which tends to reduce salinity.

The ideal salinity range for keeping marine fish in an aquarium is between 1.020 and 1.022. A slightly higher salinity level, between 1.023 and 1.025, is more suitable for keeping marine invertebrates.

With the relatively small volumes of water involved in a marine aquarium (i.e. compared to the ocean), processes such as evaporation and, to a lesser extent, loss of salt through bursting air bubbles from aeration, plus a phenomenon called 'salt creep' (in which salt is deposited around the tank and filters), can have a significant effect on the specific gravity of the water. Left unchecked, these processes result in a constantly varying salinity, which stresses the sensitive fish and invertebrates and makes them susceptible to disease. It is vital to make good these losses on a regular basis, by topping up with either fresh water or salt mix as required. A device aptly called an 'osmolator' can be beneficial in marine tanks, since it accurately monitors the water level and automatically pumps fresh water into the aquarium to replace evaporation losses and thus help maintain a constant salinity. In marine aquaria, certain vital trace elements may become depleted over time (for example some elements, such as strontium, are utilised by corals and other marine invertebrates) and so must be replaced. Commercial trace element additives are available for compensating for these salt losses.

Carbon dioxide and oxygen

The principal gases dissolved in water are carbon dioxide, oxygen and nitrogen. The relative quantities of these gases in water reflects their different solubilities. Carbon dioxide is by far the most soluble gas, while oxygen is more soluble than nitrogen; the ratio of water solubility of carbon dioxide to oxygen to nitrogen is 70:2:1.

Nitrogen gas is of relatively little significance to aquatic life, but the levels of oxygen and carbon dioxide dissolved in a body of water have a profound influence on the aquatic organisms that live within it. Carbon dioxide and oxygen are closely interlinked by the life giving processes of animal respiration and plant photosynthesis. Aerobic animal respiration requires the presence of sufficient

Above: Maintaining suitable water conditions in a marine aquarium can be a demanding task, particularly keeping the salinity at the correct level.

Optimum dissolved oxygen levels (mg/litre) for fish at different temperatures

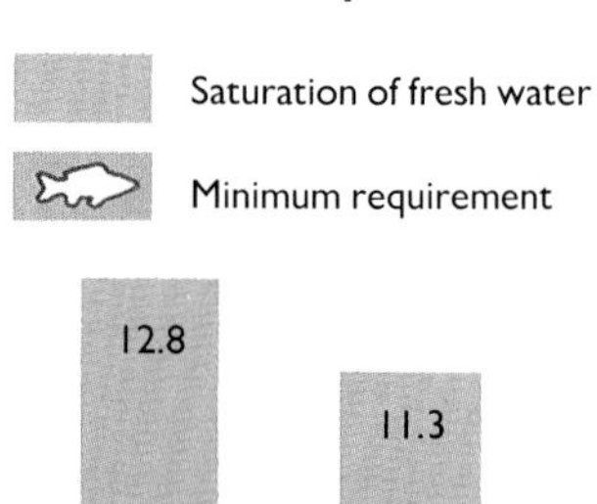

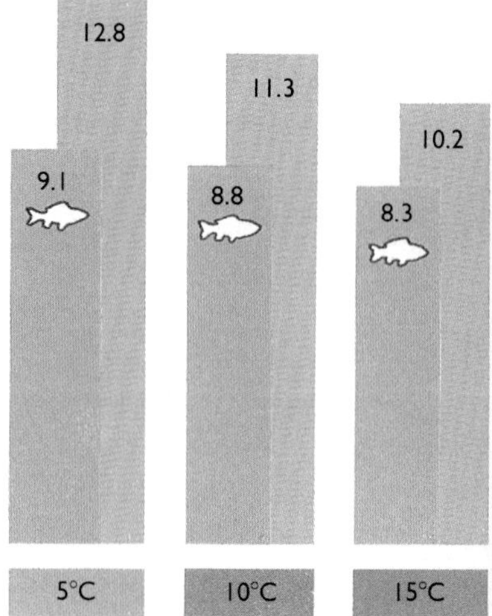

oxygen and produces carbon dioxide as a waste product, while plant photosynthesis uses carbon dioxide as a raw material and produces oxygen as a byproduct. Excessive levels of free carbon dioxide are toxic to fish and other aerobic aquatic creatures. Water's carbon dioxide content is complicated by the fact that much of it is tied up in other substances, such as bicarbonate ions (HCO_3^-).

There is 20 to 30 times less oxygen in water than in the same volume of air, which makes water a demanding environment for aerobic creatures to survive in. Oxygen solubility decreases with increasing temperature and salinity. For instance, at 10°C (50°F) the level of oxygen saturation in fresh water is 11.3mg/litre; at

Carbon dioxide and oxygen budget

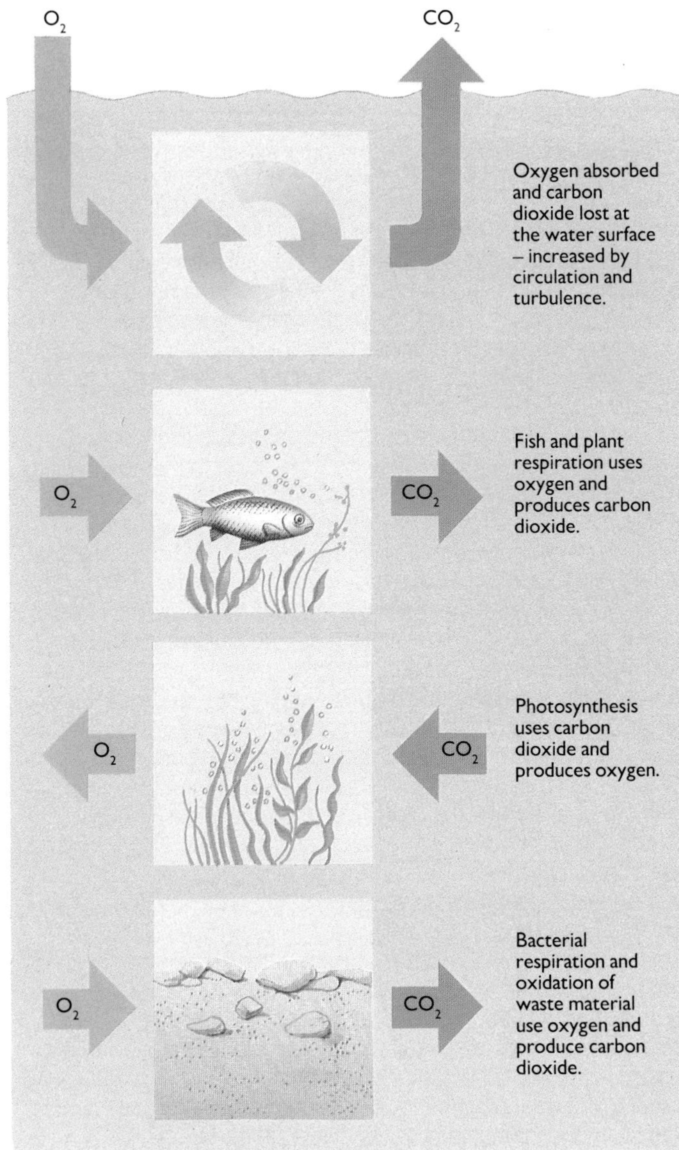

Right: How oxygen and carbon dioxide are used and produced in a typical body of water.

Below: These bar graphs highlight how the minimum oxygen demands of fish and the maximum levels available in the water edge closer as the water warms up.

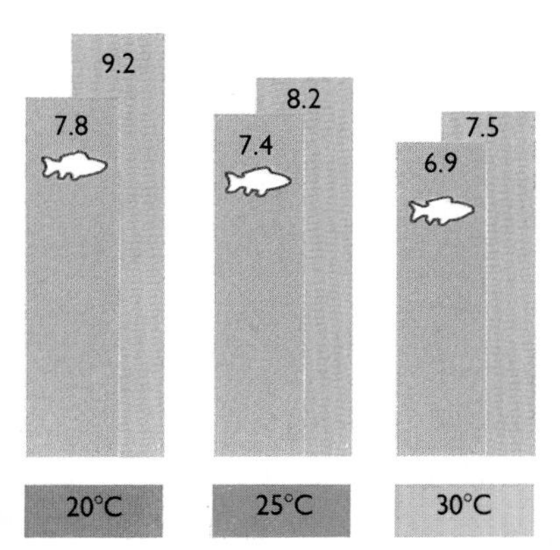

25°C(77°F), this falls by 27 percent to 8.2mg/litre, and in sea water at the same temperature the level is 30 percent lower at 4.8mg/litre.

Other influences, in addition to temperature and salinity, affect the levels of carbon dioxide and oxygen present in water. Most gas exchanges occur at the air/water interface and are influenced by the thickness of the water surface film, or laminar layer. Water turbulence caused by wind action or water movement tends to break up the laminar layer and make it thinner, thus facilitating the exchange of gases. Because there is more oxygen in the air than in the water and more carbon dioxide in water than in the air, this gaseous exchange process generally involves the water taking up free oxygen and losing free carbon dioxide, both by simple diffusion. Circulation of the water also increases the rate of gaseous exchange, since more of the water volume is brought into contact with the air.

How weight increases disproportionately with body length

These common carp figures of length/weight highlight increasing metabolic demands.

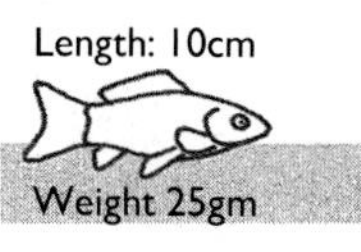

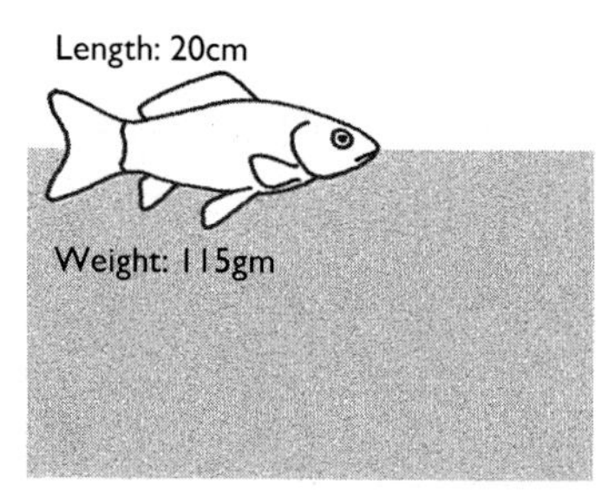

The oxygen budget

The water's oxygen content depends on the biological balance between oxygen consumption – by respiration and other oxidation processes – and the rate of replenishment by photosynthesis and diffusion at the surface. This biological balance is called the 'oxygen budget'. Quite clearly, if demand exceeds replenishment, the oxygen level falls.

Oxygen consumption is related to the number of aerobic organisms present that require oxygen for respiration, including fishes, invertebrates and plants. (Although plant respiration is cancelled out by photosynthesis during the day, it can place a considerable burden on oxygen levels at night.) The depletion of oxygen by bacterial activity can also be quite significant and depends on the organic 'loading' of the water. The more organic material the water contains the more bacterial activity is required to break it down as part of the decomposition process. This is why organic effluent or large-scale death of algal blooms can cause excessive oxygen depletion. Rotting waste also produces hydrogen sulphide (H_2S), which needs to be oxidized, thus using up oxygen. In fact, one of the important tests of water quality is its biochemical oxygen demand (BOD), which measures the amount of all organic and inorganic material that requires oxygen for its treatment. Effluent pollution containing oxygen-binding organic and inorganic compounds, i.e. ones that rapidly combine with any oxygen in the water, also has chronic oxygen-depleting consequences.

When the air/water interface is sealed by ice during the winter, decomposition and aerobic oxygen use continues but no new oxygen enters the water, causing oxygen depletion. This, coupled with a build-up of high levels of toxic gases that cannot escape, causes considerable stress to creatures living under the ice. Conversely, it is possible for excess oxygen saturation to occur when very prolific algal blooms caused by 'rich' water produce so much oxygen during photosynthesis on sunny days that up to 140 percent saturation occurs. (This means that the water becomes supersaturated with oxygen so that the gas comes out of solution. In effect, the water contains 140 percent of the oxygen that it can normally hold at that temperature.) This situation is as undesirable as oxygen depletion, as we discuss later.

How oxygen consumption changes with increasing body weight
Although total use increases, it decreases per unit body weight.

Oxygen consumption per hour (mg O_2/hr)

Oxygen consumption per gram dry body weight (mg O_2/hr/gm)

Dry body weight: 1gm

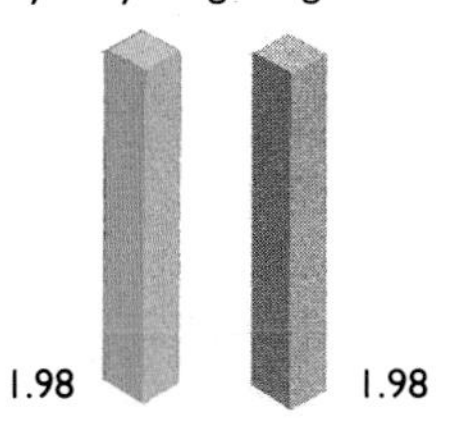

Dry body weight: 10gm

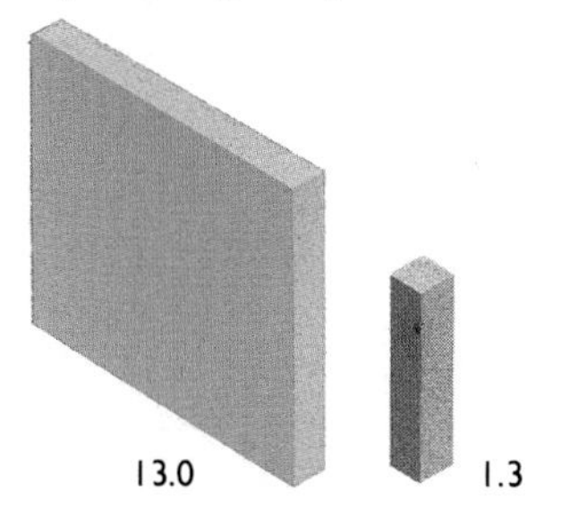

Dry body weight: 100gm

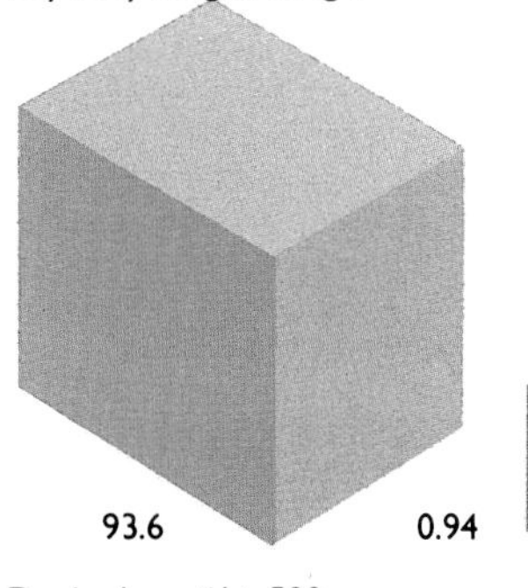

Dry body weight: 500gm

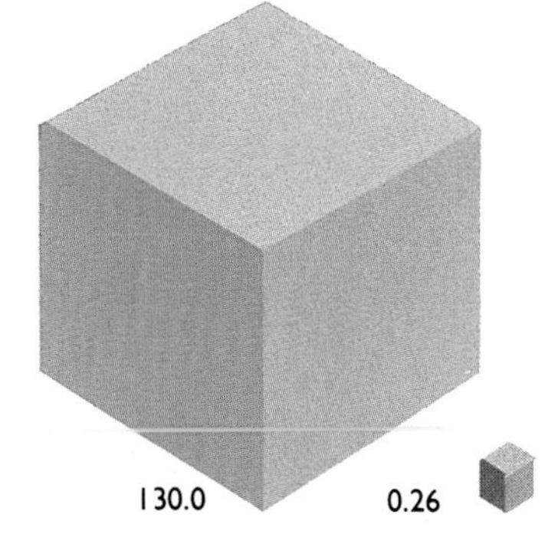

Carbon dioxide levels

The concentration of carbon dioxide in the water is decreased by aeration, turbulence and vigorous plant growth (through photosynthesis), and increased by high levels of respiration. Carbon dioxide has a complex relationship with bicarbonate, and, in equilibrium conditions, bicarbonate ions and carbon dioxide exchange freely. The equation below shows the 'balancing act' between carbon dioxide and calcium bicarbonate.

$$\underset{\text{Calcium carbonate}}{CaCO_3} + \left[\underset{\text{Water}}{H_2O} + \underset{\text{Carbon dioxide}}{CO_2} = \underset{\text{Carbonic acid}}{H_2CO_3} \right] \rightleftharpoons \underset{\text{Calcium bicarbonate}}{Ca(HCO_3)_2}$$

When there is ample carbon dioxide present, the reaction occurs from left to right. In the absence of free carbon dioxide, the reaction changes direction and goes from right to left, as bicarbonate breaks down into carbon dioxide – which is used in photosynthesis – and water. The reaction also yields insoluble carbonate – in this case, calcium carbonate.

If excess free carbon dioxide is present in hard water, the excess is mopped up by alkaline buffers, such as carbonate in the above reaction. This is not possible in soft waters, however, and so it is vital to monitor carbon dioxide levels carefully in soft water to prevent excessive build-up of the gas from respiration.

Fishes and oxygen and carbon dioxide levels

The lowest oxygen concentration required for normal functioning differs with individual species and depends on a number of physical and chemical factors, such as the fish's size (large individuals require more oxygen than small ones), the fish's age (in that metabolic rates vary with age), and the fish's physiological condition and health, especially of the gill structures. Oxygen requirements are higher for more active species; constantly active fish such as salmon (*Salmo salar*) need a minimum of 5mg of oxygen per litre of water, while some sedentary warmwater species, such as the South American leaf-fish (*Monocirrhus polyacanthus*) and the spotted talking catfish (*Agamyxis pectinifrons*), will survive on 1-2mg of oxygen per litre. Since metabolic rate is temperature related, oxygen consumption also increases at higher temperatures. For instance, carp consume 7.2mg of oxygen per kilogram of fish per hour at 2°C(36°F) and 300mg oxygen/kg/hour at 30°C(86°F).

Fish species have become adapted in both their life style and their physiology to survive in a very wide range of oxygen concentrations. The most marked adaptations are to low oxygen levels, those species that survive in these environments being generally restricted to a fairly sedentary life style. Physiologically, fishes that live in low-oxygen conditions have fine tuned their aquatic respiration system to work with the greatest possible efficiency. Dover soles (*Solea solea*), for example, extract 80 percent of the oxygen from the water passing over their gills. This is linked to their ability to tolerate very low oxygen levels in the tissues. (Since efficient extraction of oxygen by the gills depends largely on the diffusion gradient between the water and the blood, the lower the oxygen level in the blood entering the gills the more oxygen is removed from the water passing over them. Thus, a low oxygen tolerance in the tissues allows a low blood oxygen level and more efficient oxygen uptake in the gills.)

There are a number of aquatic environments in which the oxygen level periodically falls below survival levels, and so fish living in these environments have had to develop air-breathing capabilities. The majority of such air-breathing fish use this skill intermittently, thus reducing the risk of drying out and carbon dioxide diffusion problems. Carbon dioxide poses a potential problem because the gas diffuses out of a fish much more slowly in air than it does in water, quite simply because it is so soluble in water. Therefore, air-breathing fish that can remain out of water for a relatively long period, such as the aptly named 'lungfishes', have developed a higher tolerance to carbon dioxide. Air-breathing fish also show further physiological adaptations. Their ventilation rate, for example, is increased by high blood carbon dioxide levels. (Although this is a normal response in terrestrial mammals, in fish the ventilation rate is strictly related to oxygen levels.) It also appears that air-breathing fishes may be able to excrete carbon dioxide directly through the skin. In structural terms, air-breathing fish exhibit a range of solutions to the basic problem that normal fish gills collapse when out of water. Lungfish, for example, have converted the swimbladder into a simple lung structure, while the so-called 'labyrinth fishes' have an accessory respiratory organ just behind the gills. This labyrinth organ, so named because of its many-folded and blood-rich lamellae that absorb oxygen from the air, enables fishes such as Siamese fighting fishes and gouramis to survive in their native poorly oxygenated waters. Other fish, such as *Corydoras* catfish, simply swallow air and absorb oxygen through the lining of the stomach and intestine. In general, all these respiration structures share the characteristics of a large and highly convoluted surface area with a very good capillary blood supply for oxygen uptake. Some of the air-breathing fishes, such as gouramis, are obligate air breathers and must have access to the water's surface to survive.

Above: This three-spot gourami (*Trichogaster trichopterus*) has an accessory respiratory organ that enables it to 'breathe' air at the water's surface. The ability to use atmospheric air is an advantage in the poorly oxygenated waters of its native tropical environment. In fact, gouramis need access to air even when they are kept in oxygen-rich aquarium water.

Each fish species has an optimum oxygen concentration at and above which there is sufficient oxygen to support their normal life style. At low activity levels, and hence low oxygen requirements, fish have the ability to restrict the amount of blood which passes into the gills and thus reduce the contact it has with the water flowing over it. This is achieved, under hormonal control, by altering the blood's flow pattern so that it does not circulate into all the gill lamellae. This process reduces the amount of energy required for osmoregulation, because there is less active gill area for osmotic loss or gain of salt or water. When demand for oxygen increases, the blood flow is rerouted to all the gill lamellae and the gaseous exchange surface maximized once again.

If the oxygen concentration of the water falls below a fish's optimum level, this can adversely affect its growth, reproduction, activity or physiological function and, inevitably, makes the fish more susceptible to disease attack (The effects of low oxygen levels on reproduction include arrested egg development, deformities in young and high fry mortality.) If the oxygen level continues to fall to a point at which it is insufficient but respiration is still possible, a

condition known as 'hypoxia' develops. Behaviourally, at this point the fish gasps at the water's surface. Physiologically, the fish's functions adjust to the low oxygen level by decreasing the number of heart beats but increasing the heart's stroke volume, thus ensuring blood flow remains constant while consuming less energy. The ventilation volume increases; a trout, for example, 'breathes' at a rate of 70 per minute in water containing 11mg/litre of oxygen, but at 140 per minute in 3mg/litre of oxygen. If environmental oxygen levels continue to fall, at a critical minimum level breathing rate and oxygen uptake decrease because the compensation mechanism reaches breaking point, and chronic carbon dioxide blood poisoning results. At this point, the fish displays an escape response in an attempt to find higher dissolved oxygen levels. If this does not succeed, the fish slips into a coma, losing its equilibrium and rolling belly up. Fish that have died from apshyxia, i.e. total lack of oxygen, typically have flared gill covers, wide open mouths and unusually pale gills. (Rigor mortis causes similar symptoms in fish that have died from a variety of causes, but an environmental investigation should confirm whether asphyxia is the cause.) Excessive oxygen saturation, as caused by algal bloom photosynthesis or the sudden heating of cold water, can result in gas bubble disease. This is caused by excess gas in the blood vessels coming out of solution and forming small bubbles. Large numbers of these bubbles forming, especially in the gill structures, can lead to mortality.

Controlling levels of oxygen and carbon dioxide

In fishkeeping, the important aim is to keep oxygen levels high and carbon dioxide levels low. In effect, maintaining high oxygen levels and low carbon dioxide levels is achieved by the same aerating process. Those aquarists more interested in boosting aquatic plant growth may need to supplement the natural respiration production of carbon dioxide by fitting a carbon dioxide diffuser. Careful planning from the start will generally ensure that no oxygen deficit problems occur. Always follow sensible stocking guidelines for numbers of fish in any one aquarium. These are usually based on the aquarium's water surface area, which directly reflects it gaseous exchange ability. The old-fashioned belief that aquatic plants will provide enough oxygen for fishes' respiration requirements is a fallacy, since these same plants will deplete oxygen through their own respiration during the night. Gaseous exchange is improved by good water circulation and turbulence; these conditions are usually provided as a matter of course by the filtration system. Air pump-powered filters are especially effective, since the rising stream of fine bubbles provides a most effective water lifting and gaseous exchange function. It is important to note that when using water uplifts, there is an optimum air flow rate, and the belief that the more air provided the better the flow rate through the filter is not always strictly true. Many aquariums are now filtered using water pumps (power-head filters). Usually, the turbulence and circulation produced is sufficient to maintain the aquarium's oxygen levels, especially when water pumps have venturi devices fitted to improve their aeration function. However, air bubble filtration tends to result in a better removal of carbon dioxide.

Below: Clear signs of gas bubble disease. Here, the bubbles are forming in the fins and skin; such bubbles developing in delicate tissues such as the gill and eye can eventually cause death. These problems can occur when the water is supersaturated with nitrogen or oxygen. The condition is similar to 'the bends' in human divers.

When keeping some highly oxygen-sensitive species, such as marines and Rift Valley cichlids, it is sensible to supplement any water pump-powered filtration with airstone aeration. Although high-salinity water has a lower oxygen solubility, aeration is more effective because the denser medium causes the streams to be made up of smaller bubbles with a correspondingly larger surface area. This fact, coupled with the natural tendency of many dissolved organic molecules to be 'surface active' and thus 'stick' to the air/water interface (in this case the surface of the bubbles), also leads to the effective filtration technique of so-called 'protein skimming' in marine tanks, which can remove up to 80 percent of protein matter, i.e. organic waste products.

Although they have a larger surface area for gaseous exchange, garden ponds tend to pose more problems in terms of maintaining oxygen levels, even if you follow the stocking rules for plants and fish. Summer algal blooms, for example, tend to supersaturate the water with oxygen during the day – through extreme levels of photosynthesis – and deplete oxygen at night, causing early morning mortalities. Preventing algal blooms is the key; try to locate the pond out of all-day sunshine and provide a good growth of plants to compete with the algae for light and nutrients. If this fails, you may need to filter the water, both to extract the algal cells mechanically and, more importantly, to reduce the levels of nitrogenous compounds that 'fuel' the algal blooms. Installing an ultraviolet (UV) filter may sometimes help by destroying the free-swimming algae that cause green water. Low oxygen problems are particularly likely in sultry thundery weather, because air pressure is low and oxygen has a reduced solubility in warm water. Aerating the water will help, ideally by circulating the pond water through a filter and fitting a venturi on the return pipe. Failing this, waterfalls and fountains also have an aerating effect.

Above: A waterfall feature in a garden pond not only provides the pleasing sight and sound of cascading water, but also fulfils a vital aerating function for the aquatic life within the pond.

Left: Forcing air into water by using a venturi, as shown here, may be particularly useful in warm thundery weather and is widely used in koi pools to keep the water fresh and well aerated.

As we have seen on page 52, ponds freezing over in winter also cause oxygen and carbon dioxide problems, simply because the air/water interface is sealed and gaseous exchange cannot occur. Ideally, therefore, keep a small area open by installing a pond heater. Avoid overfeeding in aquatic systems, since this increases the organic load and reduces oxygen. Regular maintenance and removing dead matter will also cut down organic oxygen depletion.

Fishkeepers rarely measure oxygen levels, although kits and electronic meters are available. Generally, careful observation of the fish is adequate. Increased ventilation rate and fish gasping at the surface often indicates low oxygen levels in the water (although some gill parasites can cause these behavioural symptoms). If you suspect low oxygen conditions, it is vital to react quickly. Stop feeding to reduce organic loading of the system, apply forced aeration, and make a partial water change using well aerated water.

Ammonia

Ammonia is produced from the breakdown of protein to obtain energy and is excreted through the gills in exchange for sodium as part of the ion regulation system. Ammonia dissolved in water rapidly associates to produce ammonium ions (NH_4^+) and hydroxyl ions (OH^-). However, as the pH level rises (i.e. as the water tends to become more alkaline) and the temperature increases, so progressively more free ammonia is formed as these ions dissociate into ammonia and water (NH_3 and H_2O). Free ammonia is much more toxic to fish than ammonium ions. Thus, ammonia levels become far more critical at higher pH levels; at pH8, for example, only 5 percent of the ammonia is free, while at pH9, 20 percent of the ammonia is present in the free form. Ammonia levels also become more critical at higher temperatures; water holds five times more free ammonia at 25°C(77°F) than at 5°C(41°F). Ammonia toxicity decreases with increasing salinity of the water; ammonia is 30 percent less toxic in sea water than in fresh water of the same pH value, for example.

It is clear, therefore, that the total ammonia concentration of water can be evaluated in terms of toxicity only if the pH level, temperature and salinity of the water are known. It would be a mistake to assume that associated ammonia (NH_4^+) is not at all

Below: Using this simple ammonia test kit involves dissolving liquid reagent in the sample and interpreting the colour change against a colour scale of parts per million (ppm) nitrogen (total ammonia).

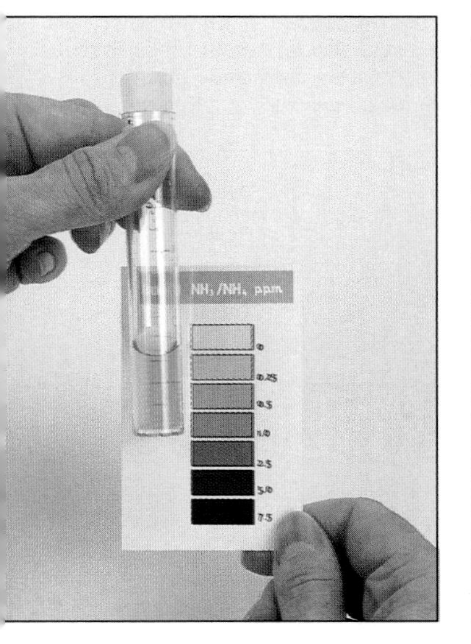

Maximum recommended level of total ammonia (as mg/litre of nitrogen indicated in widely available test kits)

pH	Water temperature 5°C	10°C	15°C	20°C	25°C
6.5	50	33.3	22.2	15.4	11.1
7.0	16.7	10.5	7.4	5.0	3.6
7.5	5.1	3.4	2.3	1.6	1.2
8.0	1.6	1.1	0.7	0.5	0.4
8.5	0.5	0.4	0.3	0.2	0.1
9.0	0.2	0.1	0.09	0.07	0.05

toxic; it is just that the free ammonia (NH_3) is considerably more toxic. Ammonia is the most toxic nitrogenous compound. It has a lowest lethal limit of toxicity for fish of 0.2–0.5mg/litre of free ammonia, i.e. this is the acute toxicity level at which fish die fairly rapidly as a direct result of ammonia poisoning. The maximum level of free ammonia that fish can tolerate for extended periods without showing chronic effects, such as increased susceptibility to disease, is 0.01–0.02mg/litre. Susceptibility to free ammonia varies with different fish species; for example, salmon and related fishes kept for a protracted period in water with as little as 0.006mg/litre of free ammonia show a chronic toxic reaction, typically gill irritation. Also, fry and eggs are more susceptible to ammonia damage than adult fishes.

The ammonia test kits available to aquarists give a reading of total ammonia, expressed as the total amount of nitrogen (mg/litre N), which reflects both the free and associated forms of ammonia. It is vital to view the result in relation to the pH value and temperature of the water to ascertain the more important figure of free ammonia. The table on page 57 shows the recommended maximum chronic exposure limit in terms of total ammonia (i.e. the figure given by the test kit) at different pH values and temperatures.

Ammonia causes a number of harmful physiological effects. It disturbs the osmoregulation system by increasing the fish's total permeability; in freshwater fishes this results in an increased urine flow, and in marine fishes an increased drinking rate. Respiration is affected because the ammonia attacks and destroys the mucus layer of the gills, causing them to swell up. This irritation also stimulates the gills to produce new cells on the surface of the lamellae – a condition known as 'hyperplasia' – which obstructs the water flow and hence reduces the oxygen uptake. Ammonia also impairs haemoglobin's ability to carry oxygen.

Lethal levels of ammonia have a generally adverse effect on fish, destroying mucous membranes of the skin and intestine, causing external bleeding and haemorrhaging of internal organs. Ammonia also damages the brain and central nervous system. At sublethal levels, ammonia is implicated in the cause of some diseases, including bacterial gill disease (in which the production of hyperplasic cells increases the vulnerability of the gills to bacterial invasion), dropsy and finrot.

Above: The effects of drastic ammonia poisoning are clear in this fancy goldfish, with the structure of the skin visibly breaking down along the back.

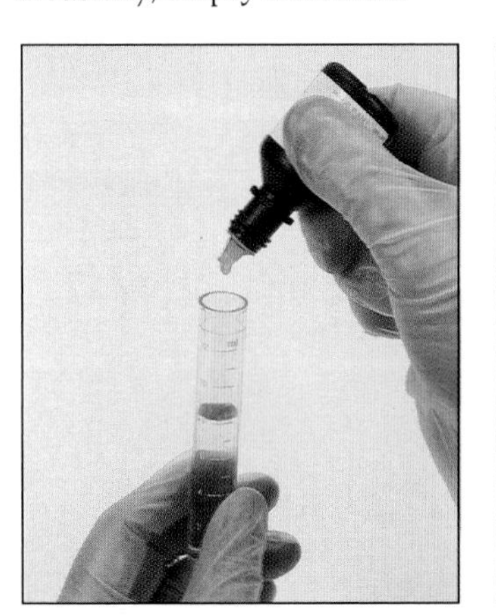

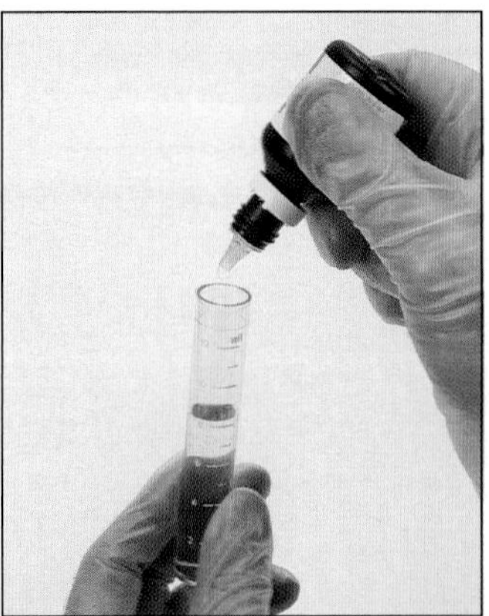

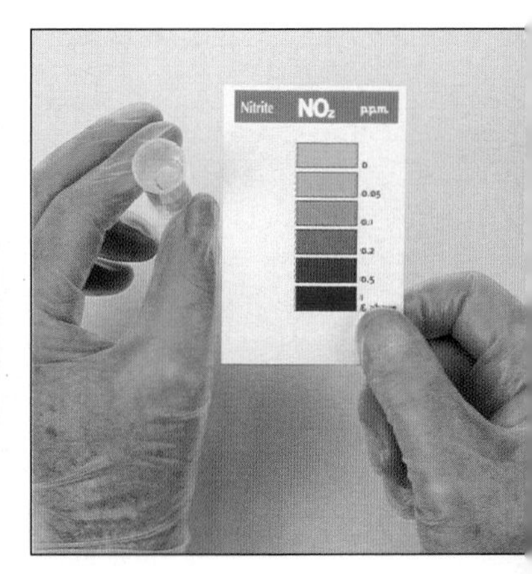

Below: Testing for nitrite level entails adding one and then another reagent to a sample of water. Matching the colour change with a colour reference card provides a readout of nitrite in parts per million (ppm). Simple dip-stick test kits are also available for detecting nitrite and other water parameters.

Nitrite

In the presence of oxygen, ammonia is converted into nitrite (NO_2^-) by bacteria of the *Nitrosomonas* species. This is one of the steps in the process called nitrification. Nitrite is less toxic than ammonia, being lethal at levels of 10-20mg/litre. Again, this lethal toxicity level varies with species. Guppies, for example, can cope with levels up to 100mg/litre before it proves lethal, while some fish, such as discus (*Symphysodon discus*), suffer increased disease susceptibility at levels as low as 0.5mg/litre.

Nitrite proves toxic because it breaks down the red blood cells and oxidizes the iron in the haemoglobin into a stable state called methaemoglobin, which has no oxygen-carrying capacity. (This process has the effect of turning the gills and blood brown.) The ability to convert methaemoglobin back into haemoglobin determines how resistant a particular fish is to high nitrite levels.

The nitrogen cycle
A simplified view of how nitrogen circulates in the natural world.

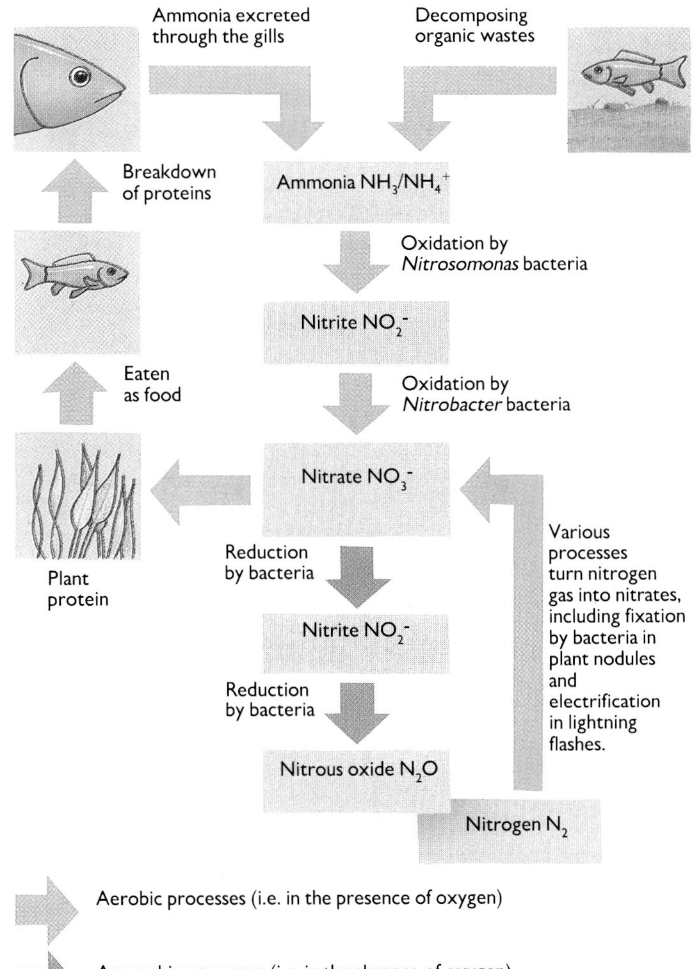

Right: At the heart of biological filtration processes lies the nitrogen cycle, a natural and ever-turning sequence of biochemical reactions that underpins any discussion about maintaining water quality. This illustration is a simplified representation of the stages involved. The crucial stages for the fishkeeper are the conversions of toxic ammonia to nitrite and then nitrate. These vital processes are carried out by nitrifying bacteria that live in well-oxygenated conditions. Anaerobic bacteria turn nitrate into nitrogen.

Nitrite poisoning symptoms are listlessness and anoxia (oxygen starvation caused by the level falling below that necessary for respiration to continue, i.e. below that of hypoxia – see pages 54-55) and pigmentation (typically in the form of dark spots) of the liver, spleen and kidney.

Increased dissolved salt content reduces the toxicity of nitrite to fish. Thus, nitrite is less toxic in sea water and has reduced toxicity in hard water. For example, the same toxic effect has been noted at 18mg/litre of nitrite in hard water and at 10mg/litre in soft water. The mechanism for this has not definitely been proven, but whatever the explanation, adding salt (sodium chloride) to a concentration of 0.01 percent (0.1g per litre or approximately 1 level teaspoon per 10 gallons) does reduce the stress effect of high nitrite levels in fresh water. Most freshwater fish will tolerate this low level of salt.

Nitrate

The nitrification process continues as *Nitrobacter* species bacteria oxidize nitrite into the even less toxic nitrate ion (NO_3^-). Nitrate has lowest limits of lethal toxicity of 50-300mg/litre, and this is with extremely sensitive fish. Nitrate is 2000 times less toxic to trout, for example, than nitrite, but fish eggs and fry are much more sensitive to nitrate in the water than adult fish. Nitrate has been found to be more toxic in salt water than in fresh water. It is reasonable to assume that very high levels of nitrate cause stress and greater susceptibility to disease in marine aquariums since nitrate levels in the oceans are almost zero. Certainly, real success in keeping delicate marine invertebrates, which are generally more sensitive than fish, has resulted only from keeping nitrate levels below 20mg/litre. Controlling nitrate levels has also proved more critical for keeping discus and Rift Valley cichlid communities.

Controlling levels of ammonia, nitrite and nitrate

In the 'closed' environment of an aquarium or pond, nitrogenous waste in the form of ammonia can accumulate and reach toxic levels if nothing is done to counter it. Breaking ammonia down to less toxic substances is the major role of biological filtration. Although there are many different types of biological filters, the overall working principles are the same. Biological filtration provides the best possible conditions to encourage the nitrifying process, by which specific bacteria break ammonia down in two stages into the progressively less toxic nitrite and then nitrate. All biological filters contain media with the largest possible surface area for nitrifying bacteria to colonize. These bacteria require a constant stream of nutrients in the form of nitrogenous compounds and a generous supply of oxygen to carry out the oxidation process. Both of these conditions are provided by circulating well-oxygenated water rich in nitrogenous waste through the filter medium. The nitrification process is also pH and temperature dependent, working at its most efficient at pH 7.5 and relatively high temperatures (30°C/86°F in fresh and 30-35°C/86-95°F in sea water), which are generally too high for most fishkeeping. It is easy to monitor ammonia, nitrite and nitrate levels using very simple test kits. Ideally, keep the total ammonia and nitrite levels

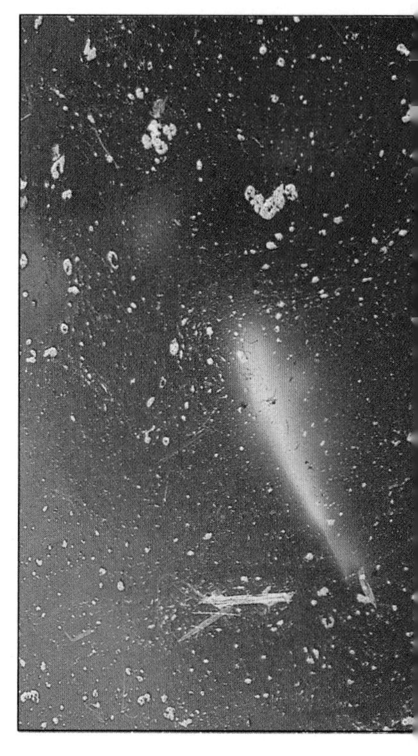

Above: The scum floating on the surface of this koi pool indicates that there is an excess of protein waste. Efficient biological filtration will clear the water.

Below: Taking regular tests of nitrate level, here registered as nitrate-nitrogen in mg/litre, is vital for keeping delicate species such as marine invertebrates.

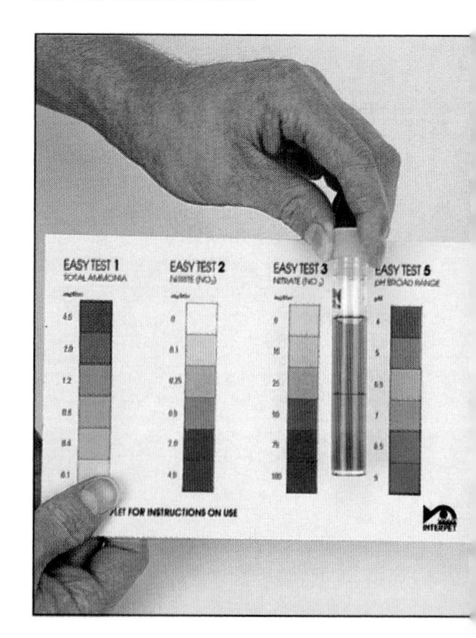

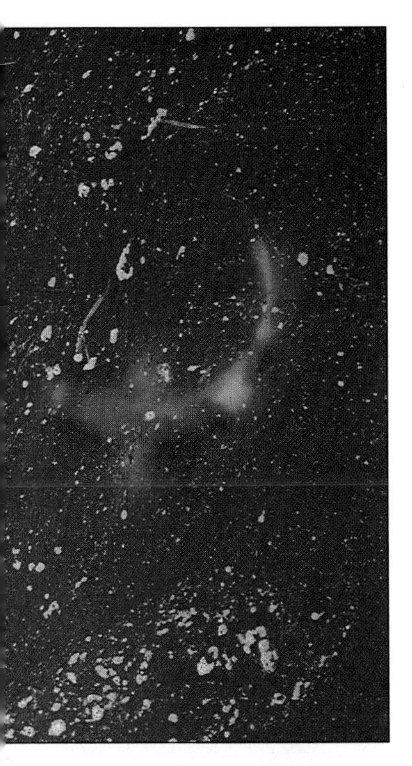

below 0.1mg/litre, and nitrate below 20mg/litre. An efficient filtration system should keep the ammonia and nitrite levels well below this limit. If the concentrations rise above this level, there are a number of possible causes. One of the most likely explanations is that the filter has not fully matured – a condition known as the 'new tank syndrome'. At 25°C (77°F), a biological filter will take about two to six weeks to develop a full complement of nitrifying bacteria; at 10°C (50°F), this process of maturation can take four to eight weeks. It is possible to speed up the maturation process by seeding the filter with bacteria, either in the form of a freeze-dried or liquid commercial preparation or by adding some filter medium from a mature tank (provided the donor tank has no recent history of disease problems).

In fact, a filter may take six months to stabilize, i.e. to reach the point at which it contains a balanced population of nitrifying bacteria. Therefore, to be on the safe side, it is advisable not to stock your body of water to its maximum limit of fish until after six months. Remember that the size of the bacterial population in a fully mature filter is directly proportional to the amount of nitrogenous 'food' provided by the established fish population. Therefore, take care to build up the number of fish in an aquarium or pond *gradually* over a period of time to allow the bacteria to increase in step so that they can cope with the escalating quantities of waste products. As a guide, add just a couple of fish initially and monitor ammonia and nitrite levels closely. Wait 2–3 weeks before adding a few more fish, and only if ammonia and nitrite readings remain near zero. If a high ammonia or nitrite reading is obtained perform a partial water change (up to 50 percent) immediately. Undertake additional water changes if necessary, to reduce ammonia and nitrite concentrations to safe levels.

Below: In a newly set-up pond or aquarium, the levels of ammonia and nitrite form overlapping peaks as the bacteria in the biological filter build up in numbers.

The maturing process of a biological filter

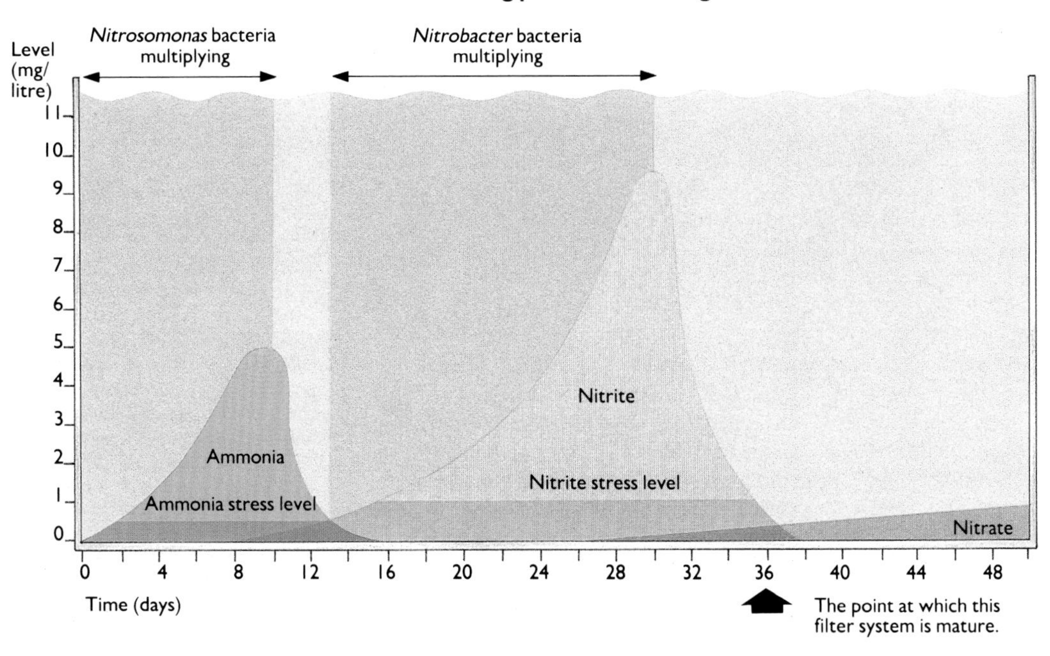

High ammonia and nitrite levels may also indicate that the filter is overloaded, either because the fish population is too large or the filter bed is too small. In both cases, there will not be enough bacteria in the biological filter to cope with the nitrogenous load. Overfeeding can also have this overloading effect, since any waste food will disintegrate and be broken down into ammonia. Providing a suitable filtration system, and carefully following stocking guidelines, as well as sensible feeding regimes, will ensure that this problem does not occur. Another cause of high ammonia or nitrite is excessive cleaning of the filter medium, such that too many filter bacteria are washed away, so reducing the filter's ability to cope with the fish's wastes. To avoid this, gently rinse the filter medium in some water taken from the aquarium or pond. Tap water should not be used as it contains chlorine that will harm the filter bacteria.

If all other possibilities have been discounted, high ammonia and nitrate may indicate that the filter bacteria have been inhibited or destroyed by some type of toxin. Bacterial filter toxins usually take the form of synthetic chemicals, such as insecticides, herbicides or household sprays, but fish medications, such as methylene blue and some antibiotics, may also have the same effect. Always try to prevent any likely toxins coming into contact with the filter bed, but if they do, remove them rapidly by filtering the water through activated carbon and/or a synthetic chemical adsorbing pad and making partial water changes.

To actively adsorb excess ammonia, use a chemical filtration medium, such as zeolite (a natural substance) or a synthetic adsorption pad, and carry out partial water changes of up to 75 percent, depending on the severity of the problem. (Zeolite will not work in sea water; synthetic adsorption pads will.) Take care that the pH value of the aquarium is not suddenly raised by these water changes; as we have seen, more free ammonia is present at higher pH levels and hence the overall toxicity to the fish increases. To counteract high nitrite levels, make partial water changes of up to 50 percent and add salt (sodium chloride) to a concentration of 0.01 percent to reduce the stressful effects of the nitrite. In both these cases, try to discover and rectify the cause as soon as possible.

Accumulating nitrate can be controlled by routine management; a fortnightly 25 percent water change will dilute and lower the nitrate level, for example. (Do check the nitrate level in your tapwater first, however. If it is unduly high, filter the water through a deionizing resin – some are available to remove nitrate specifically – or use rainwater or RO water – see page 48). Vigorous plant growth also removes nitrate, since it is an essential plant food. There are also some denitrifying filtration techniques available that provide oxygen-starved conditions, i.e. below 1mg/litre of oxygen, in which certain anaerobic bacteria must obtain their oxygen supply by removing it from the nitrate (NO_3^-) ions, finally leaving pure nitrogen gas as a waste product. These anaerobic conditions are provided either in trickle columns, anaerobic boxes or in the microporous structure of certain filter media. Denitrification also requires a carbon source to 'feed' the bacteria, and this can be provided either in the media structure or as a separate liquid or solid 'food'.

Above: The aquatic plants in a mixed community aquarium will absorb a certain amount of the nitrates as a 'food source', but regular partial water changes are still necessary to keep nitrates down to suitable levels.

Right: To reduce the levels of chlorine, be sure to treat raw tapwater with a conditioner. This simply involves adding the required amount to the water as specified by the manufacturer. Some water conditioners will also remove chloramine, a more stable form of disinfectant that is sometimes added to tapwater. Be sure to pre-treat the tapwater with water conditioner (e.g. in a clean bucket) *before* adding it to the aquarium.

Chlorine and chloramine

Most water supplies are dosed with chlorine as a disinfectant measure. The chlorine is forced into water as a gas, the high concentration of chlorine driving off other gases. A small proportion remains as so-called 'free chlorine', but most combines with water until an equilibrium is reached, as shown in the equation below.

$$Cl_2 + H_2O \rightleftharpoons HOCl + H^+ + Cl^-$$

Cl_2	H_2O	HOCl	H^+	Cl^-
Free chlorine	Water	Hypochlorous acid	Hydrogen ion	Chlorine ion

Depending on the temperature and pH value of the water, the chlorine that combines with water may be present in its associated form of hypochlorous acid, as shown above, or dissociated as ions of hydrogen and hypochlorite, as shown below.

$$HOCl \underset{\text{Acid pH and increasing temperature}}{\overset{\text{Alkaline pH and falling temperature}}{\rightleftharpoons}} H^+ + OCl^-$$

HOCl		H^+	OCl^-
Hypochlorous acid		Hydrogen ion	Hypochlorite ion

It is the hypochlorous acid, rather than the dissociated hypochlorite ions, that acts as the toxin/disinfectant. It is thought to chlorinate cell proteins and enzyme systems and thus is extremely toxic to fish. Chlorine is relatively unstable in water – more so at higher temperatures – and its disinfectant properties are reduced as it reacts with organic matter and other substances and as a gas it also diffuses into the air, where chlorine concentrations are much lower. (Fishkeepers can usefully speed up this diffusion process by aerating the water vigorously.) Water supply companies, however, are principally interested in extending chlorine's disinfecting properties, and this can be a problem where there are long pipelines between the water treatment plant and the consumers. The use of chlorine disinfectant in long pipelines involves the operator having to calculate the fall off in chlorine levels over the full length of the pipe. The dosage level has to be sufficient to achieve the recommended residual chlorine level of 0.2–0.5mg/litre at the far end of the pipeline. Thus, to compensate for losses along the way, the residual chlorine level present in the water supply at the beginning of the pipeline is considerably higher than 0.2–0.5mg/litre. Another solution has been to combine chlorine with nitrogenous compounds, such as ammonia, to form chloramine. This is much more stable than free chlorine, releasing hypochlorous acid more slowly and in twice the quantity per atom of chlorine.

Since chloramine remains more stable if an excess of ammonia is present in the water, supply companies tend to add more ammonia than is necessary. Any excess ammonia in the water obviously poses a threat to fish, as we have seen on page 58. Water supply companies also add excess free chlorine and chloramine to replace hypochlorous acid lost through disinfection. Chloramines can also be formed when naturally occurring nitrogenous compounds – nitrate fertilizers, for example – enter chlorinated water supplies.

Chlorine in the water causes an 'escape response', as affected fish attempt to find chlorine-free water. If escape is impossible, the fishes begin to tremble and discolour, becoming listless and weak. Eventually, they stop ventilating because of the chlorine's destructive effect on their respiratory tissue. The level of chlorine that proves lethal to fish depends on: the water's pH level and temperature (which, as we have seen, dictates the proportion of hypochlorous acid present); the amount of residual chlorine present (which is the total free chlorine and chloramine content of the water and also dictates the quantity of hypochlorous acid produced); the presence of some polluting chemicals (they can increase the toxic effect of chlorine); the amount of dissolved oxygen present (i.e. in relation to the degree of damage suffered by the respiratory tissue of the fish); and finally on the species (some are more susceptible than others). Chlorine toxicity levels are usually quoted in terms of residual chlorine, 0.2-0.3mg/litre being sufficient to kill most fish fairly rapidly. To avoid chronic toxic effects, the residual chlorine level should not exceed 0.003mg/litre.

It is fairly easy to avoid chlorine problems by treating raw tapwater with a suitable water conditioner. Most commercial water conditioners are based on sodium thiosulphate, which very rapidly detoxifies chlorine by chemically binding it. Water containing chloramine must be treated with a special water conditioner. This contains sodium thiosulphate to neutralize the chlorine component of chloramine plus a reagent that mops up ammonia released by the degraded chloramine molecules. To find out whether there is chloramine in your local water supply, either ask your local water supply company or use one of the widely available and easy-to-use test kits. If in doubt, use a water conditioner that removes both chlorine and chloramine.

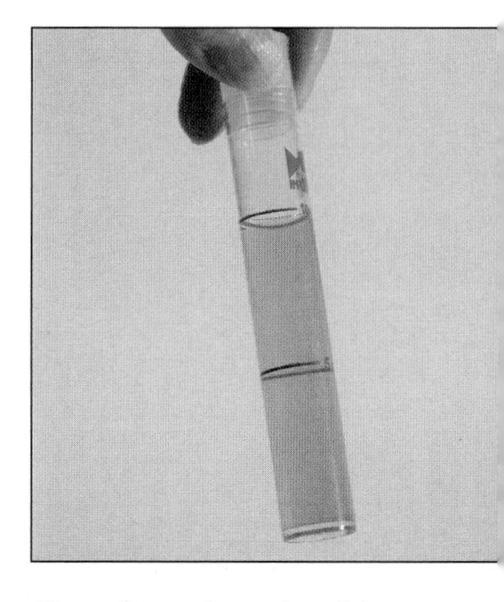

Above: Copper is very harmful to aquatic life. Use test kits such as this to monitor levels in the water, particularly to prevent overdosing when using remedies containing copper compounds.

Metal toxicity

Some water supplies contain metals, such as iron, lead and copper. These either enter the supply from source or are picked up from contact with metal pipes. This is particularly the case with soft water, which tends to be more corrosive and in which metals are more soluble. Sea water is particularly corrosive to metals, and so to avoid toxicity problems, do not use any metal (such as retaining clips) in a marine aquarium.

Metals in water can exist as several chemical 'forms', each of which has a different toxicity to fish. The form present depends on the water hardness, pH value, temperature and other dissolved substances in the water. For example, copper is more soluble in soft water, in which it exists as highly toxic free copper, while in hard water it forms copper carbonate, which precipitates out and is far less toxic to fish. Different metals also have different toxicity levels for various fish species and are more toxic in combinations than on their own. Iron and lead, for example, should never be present in levels above 0.03mg/litre for complete safety, whereas copper levels should be half that, i.e. 0.015mg/litre. You can monitor copper levels using a simple test kit.

Diagnosing lethal doses of heavy metals is extremely difficult, since it involves heavy metal analysis of fish tissue and microscopic investigation of pathological changes in the organs. (Most metals

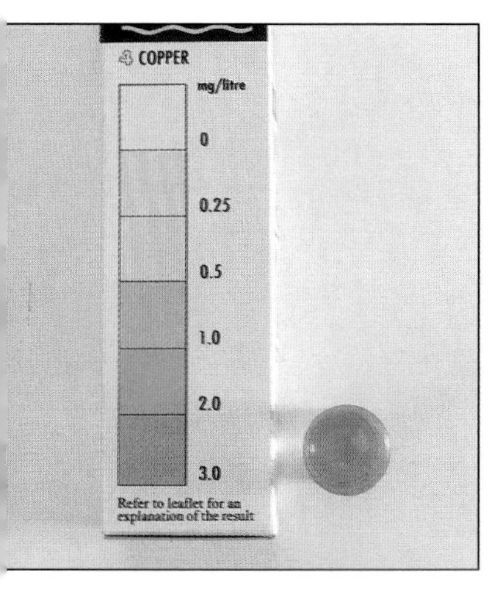

Below: Fish die in the wild from the same range of causes as those in captivity. Perhaps this fish has succumbed to a parasite infection or to high levels of pollution in the water of its native stream.

damage the blood, internal organs and the gill membranes.) For sudden unexplained deaths, particularly in softwater areas, be sure to suspect heavy metal toxicity as a possible cause. As a general precaution, always allow tapwater to run for five to ten minutes to flush out any accumulated metals. (Follow the same precaution with plastic hoses to flush out toxic plasticizers, which may be released in water standing in contact with the hose).

Pesticides and water quality

Water supply companies sometimes add insecticides to the water to kill off any pests, such as water lice (*Asellus aquaticus*) living in the water supply systems. These insecticides may take the form of pyrethrins, or the man-made equivalent, permethrin. These are usually added to the water supply in spring and autumn at levels of 5-10 micrograms/litre (i.e. 5-10 millionths of a gram per litre) over a seven-day period. Although these are very low concentrations, they can prove toxic to some fish, and water companies suggest that the water supply should not be used for fishkeeping purposes for a 14-day period from the first day of insecticide dosage. Water companies generally post warnings of insecticide treatment in the local press and are usually quite happy to supply aquarium shops and clubs with details on request.

Susceptibility to pyrethrin and permethrin depends upon species; killifishes, for example, die within 48 hours at levels of 74 micrograms/litre, while trout will die within 48 hours at levels as low as 2.5-6 micrograms/litre. It is a sobering thought that the only real safeguard against pyrethrin fish kills is to leave water to stand for one month before using it for fishkeeping. During this period, the pyrethrins break down naturally. On the domestic scene, never use insecticides, herbicides, household sprays or potentially toxic paints or varnishes anywhere near fish tanks or ponds. Also, do not use toxic substances near air pump inlets if they are located in another room or remote from the tank.

Medications as toxic agents

Many medications are toxic not only to the disease organisms but also, at higher concentrations, to the fish itself. Copper, for example, is particularly toxic to fish and especially to marine invertebrates. The toxicity of medications varies with species; labeos (such as the red-tailed black shark), for example, are particularly sensitive to copper, which at quite low concentrations attacks their skin and gills. The toxicity of medications also varies with the water's physical and chemical characteristics; those containing copper, for example, will be affected as described under *Metal toxicity*. To be safe, always use a reliable test kit to monitor the concentration of copper in the water when using a copper medication.

Although it is possible to make up and use your own medications, you need to take the greatest care and carry out careful research beforehand. It is always much safer to use commercially prepared medications, since they will have been thoroughly researched and tested in a range of conditions and with a range of species. They will also have a built-in safety margin to allow for any mistakes in the dosing procedure. (See Chapter 7, starting on page 182, for more guidance on using treatments.)

CHAPTER 4

PLANNING FOR HEALTH

Having considered the form and function of fish and the chemistry of their watery environment, our attention now turns to the practical aspects of providing healthy conditions for pond and aquarium fish. The first step is to gather as much information as possible about the natural environment and behavioural characteristics of the fishes you wish to keep. The more the conditions you provide differ from those in the wild, the more likely the fishes are to be stressed and susceptible to disease. A little forethought in planning and subsequent regular maintenance can go a long way towards keeping your fishes healthy.

Sensible stocking levels

Many fishkeepers tend to overstock their aquarium or pond. Maximum safe stocking levels are shown on pages 69 and 71. These figures assume small fishes of 2.5–5cm (1-2in) in length and allow a certain amount of room for subsequent growth, plus a margin for technical problems, such as pump failure. These figures are only a guide, since very active fish need more space than sedentary species, and installing different aeration and filtration systems can have a radical effect on total fish capacity.

The size of the aquarium or pond may be influenced strongly by the cost and available space, and this in turn influences the type and number of fishes you can keep. Although many fishes are quite happy with a solitary existence, most fishes normally maintained in tropical community aquariums are best kept as pairs or in shoals of five to ten individuals. If you wish to keep shoaling fishes, always ensure that the aquarium is sufficiently spacious for them to form natural shoals; in cramped surroundings such fishes will swim at random and suffer increased stress.

Compatibility

When setting up an aquatic community, be it aquarium or pond, be sure to choose fishes that are compatible, not only in relation to water chemistry and temperature, but also in their size, behaviour and eating habits. For example, although oscars (*Astronotus ocellatus*) and neon tetras (*Paracheirodon innesi*) both thrive in soft water, when the oscars exceed about 2.5cm (1in) long they will simply regard the neons as a tasty snack. This example highlights

Left: These angelfishes (*Pterophyllum scalare*) thrive in soft, slightly acid water and in a deep, peaceful tank. Providing these conditions allows the fishes to show their elegance to the full.

the point that small cute fish can develop into very large hungry ones. Similarly, although neons and sphenops mollies (*Poecilia sphenops*) are reasonably compatible in size, the mollies flourish best in hard alkaline water, which would not suit neons at all.

A very common mistake is to keep a shoal of fin-nipping fishes, such as black widows (*Gymnocorymbus ternetzi*) or tiger barbs (*Barbus tetrazona*), in the same aquarium as long-finned fishes, such as angelfish (*Pterophyllum scalare*). Not only will the boisterous, fast-moving fishes unsettle the relatively sedate angelfishes, and thus make them vulnerable to various stress-related diseases, but also any damage to their fins can soon lead to severe outbreaks of finrot and other infections.

Position and decor

Positioning is important. Although early morning light can act as a stimulus for fish to breed, in general, aquariums should be kept out of too much sunlight. Ponds usually flourish best in a position without too much shade. Some fishes, both in aquariums and ponds, are more timid than others and it is wise to avoid positions where there is excessive noise or movement. Locating an aquarium close to a door is not a good idea, since constant banging can stress the fish and any draughts can cause unnecessary temperature fluctuations.

The decor will depend very much on the fishes you are keeping. In general, furnishing a community aquarium with plenty of plants, rocks, caves and, possibly, flowerpots on their sides will provide refuges for timid fishes and breeding sites for others. Sometimes, the markings on the fishes give a clue to the most suitable decor. Wild-caught angelfishes, for example, have vertical markings and appreciate tall, rather straight plants, such as *Vallisneria*.

Filtration and aeration

Filtration and aeration systems carry out four major functions: to mechanically remove any debris floating in the water; to act as a chemical or biological system for eliminating toxic materials; to ensure that there is adequate exchange of gases so that oxygen can enter the water and carbon dioxide and other gases can escape; and to provide adequate circulation to give some form of natural current appropriate for the species of fish being kept.

The most suitable filter depends on the following factors:

● **The amount of sediment produced.** Goldfishes and large cichlids, for example, produce considerable quantities of waste and need a high-capacity filter to remove it. However, small characins produce very little waste and are therefore well served by a filtration system that circulates and aerates but does not require a high-capacity filter box.

● **The amount of circulation required.** Many fishkeepers equate filtration efficiency with a high circulation rate, but a relatively slow filter can be very effective, while a high-turnover filter may not be so efficient. Fishes not accustomed to fast-running streams may not have the physique for dealing with powerful currents and, even if they are not physically damaged, they may succumb through too much stress. However, even a slight breeze on a static pond in the

Aquariums

● Use lime-free gravel in softwater tanks. Avoid using coral sand or other marine media in freshwater tanks, since these are high in calcium and will harden the water.

● Try not to keep hardwater and softwater species together. Also, if using salt in the aquarium, such as with mollies, check that all species are salt tolerant.

● Place the tank away from direct sunlight, which can cause overheating and excessive algae growth. If possible, arrange for artificial lighting to turn on and off gradually. This is particularly important when breeding fish that care for their young.

● Take into account the territorial habits of some fishes when planning the aquarium layout. One pair of large cichlids, for example, needs a good deal of space. With territorial fishes, it is important to get the female used to the aquarium first before introducing the male.

● Use a variety of foods but avoid overfeeding. Most fishes will thrive on high-quality prepared foods, but appreciate a supplement of natural foods. For example, herbivore fish may benefit from occasional fresh vegetable supplements, such as a small blanched lettuce leaf, a few peas, or a slice of cucumber. Whereas carnivore fish may enjoy small pieces of fish or shellfish (e.g. prawn). Take care not to introduce disease when feeding live foods of natural origin. Live *Tubifex* worms probably pose the biggest potential source of disease. Frozen irradiated 'live' foods (sold in foil or blister packs), such as frozen daphnia, tubifex, mosquito larvae, and bloodworms, are much safer to feed.

● Carry out regular partial water changes. Frequency will depend upon the stocking level and sophistication of your filtration system. In freshwater and marine aquariums, remove about 25 percent of the tank volume every 2-4 weeks and top

up with conditioned water of the same quality and temperature.

● Do not overstock. The recommended levels shown in the table are for small fishes up to 5cm(2in). Allow more space for larger fishes, since their body weight is greater in proportion to their length (see page 52). Also allow space for growth.

● Before buying a fish, find out its adult size. Most fish are sold as juveniles and some species grow very large. Unless you can comfortably accommodate the fish when fully grown, don't buy it.

● In marine tanks, synthetic sea water prepared using a reliable marine salt mix is preferable to using natural sea water, which may be polluted. A specific gravity of 1.020-1.022 is suitable for most tropical marine life.

● Check the water regularly for temperature, pH, hardness, ammonia, nitrite, nitrate and – in marine aquariums – specific gravity (salinity).

● Avoid sudden temperature changes; 23-26°C(73-79°F) is suitable for most tropical species in the aquarium.

● Avoid sudden changes in pH value. 6.5-7.5 is fine for most freshwater tropical aquarium fish; 7.9-8.3 is satisfactory for most marine aquarium fish.

● Avoid extreme hardness values for most freshwater fish. A carbonate hardness (KH) value of 90-125 mg/litre $CaCO_3$ (5-7°dH) ensures adequate buffering in a marine aquarium.

● Ideally, keep ammonia and nitrite at negligible levels in established aquariums. Keep nitrate less than 20mg/litre for delicate marine fish and invertebrates.

● In most aquariums provide 15-20 watts of cool white fluorescent lighting or 30-40 watts of Grolux lighting per 900cm^2 (1ft^2) of surface area; more if the tank is deeper than 45cm(18in). Marine aquariums with corals, anemones and growths of 'leaf' algae need high-intensity lighting.

Type of fish (Aquarium)	Water surface area per unit body length of fish (exluding tail fin)	
	1 cm per	1 inch per
Coldwater and tropical marines	120cm^2	48in^2
Cold freshwater	75cm^2	30in^2
Tropical freshwater	25cm^2	10in^2

wild may produce considerable currents, so a compromise must be achieved. In a freshwater aquarium or pond a typical slow-running filter may turn over the volume of water in 1-2 hours, while more powerful systems in freshwater and, especially, marine aquariums commonly produce a water turnover of 1-5 times per hour.

● **The oxygen requirement.** The more crowded and warmer the aquarium, the more aeration will be needed. Some fishes are more sensitive to oxygen content than others; orfe, for example, may succumb to oxygen shortage in warm water. Air-operated filters clearly help to increase the oxygen content of the water, and some motor filters ('power filters') have either a venturi device for incorporating air into the returned water stream or a spraybar system to break up the surface of the water. Bringing the aquarium water into contact with the air has a secondary function; as oxygen is diffusing into the water, waste gases, such as carbon dioxide, have the opportunity to escape.

● **The choice of filtration medium.** It is important to select the most appropriate filter design for the medium (and therefore the type of filtration) required. Most filters are designed to work biologically, containing media that will support colonies of nitrifying bacteria. Certain filters (e.g. some canister filters) can accommodate a variety of media, including biological, mechanical, and chemical types. Less common are filters designed to harbour denitrfying bacteria.

Introducing fishes

It frequently happens that fishes which are apparently healthy in the dealers' tanks develop diseases, such as white spot, due to the stress caused in moving them to a fresh environment. It makes good sense to check the pH, hardness and, where applicable, salinity levels in the tank before introducing new fishes. It is important, therefore, to minimize this stress as much as possible. Fish are normally supplied in plastic bags and it is vital to keep these well wrapped during transit to avoid bright light and temperature fluctuations. On arrival home, float each bag in the aquarium for 15 minutes to equalize the temperature with that of the aquarium. Keep the aquarium lights switched off during this period, so as not to stress or overheat the bagged fish. In the case of pond fish, ensure the bag does not drift into direct sunlight. Then gently release the fish making sure they are disturbed as little as possible. Keep a close check on the new fish for the first few days after introduction to ensure they have settled in and are feeding.

Quarantine

Quarantining all new fish can be very important in preventing the introduction of disease into set-up aquariums and ponds. Apparently healthy fish can carry a huge range of disease organisms as low-level infections. These infections are often extremely difficult to detect, yet may wreak havoc when introduced into the relatively overcrowded confines of a pond or aquarium.

Ponds

● Ensure that the pond is deep enough for the fishes to withstand the severest winter in your region, i.e. without ice reaching the bottom. Generally, 45-60cm(18-24in) is a minimum.

● Place the pond where there is a small amount of shade but avoid areas where leaves will collect excessively in the water during the autumn. Avoid siting it where there would be undesirable run-off into the water or near a frequently used path that might cause disturbance.

● If the pond is frequented by cats, herons or other potential fish-predators then it is best to cover it with pond netting, or erect a barrier around its perimeter. Alternatively, try one of the harmless cat and heron scarers on the market.

● Most ponds benefit from thoughtful use of submerged and marginal plants and lilies. Fishes will use these as refuges and those plants that provide shade will help to keep the water clear of excessive algae.

● Do not overstock a pond. For small fishes allow $120 cm^2$ of surface area per cm of fish length, excluding the tail fin ($48 in^2$ per in). Remember that large fish need more space and that small fish will grow into large fish!

● Check the pH value of the water regularly (as shown on pages 38 and 39). An ideal range is pH 6.8-8.2. If the water becomes too acid or too alkaline, use an appropriate proprietary pH adjuster as instructed by the manufacturer.

● If the water turns green, this may be caused by too much light and/or excessive amounts of nutrients in the water fuelling an algal bloom. The fish probably do not mind it as much as you do! Watch out for signs of de-oxygenation in the early morning and if the algal bloom dies back suddenly. It is often possible to clear green water by introducing a bucketful of water from a clean healthy pond; the organisms introduced feed on the algae.

● A fountain or waterfall helps to aerate the water, which is particularly beneficial in warm thundery conditions. Remember to turn it off in freezing weather.

● Some fish treatments will adversely affect the nitrifying bacteria in a biological pond filter. Therefore, if possible, isolate the filter before treating the pond. Alternatively, remove part of the filter medium and return it at the end of the treatment.

● Long-finned fishes, such as veiltail goldfishes, suffer from fin congestion when the water temperature falls below about 10°C(50°F) and this can lead to finrot. Therefore, these fishes are not suitable for overwintering in a pond that may freeze.

● In preparation for the winter, feed the fishes well, cut back excessive plants and ensure that the pond is quite clean. Many fishes are killed through toxic gases developing under the ice. Consider installing a pond heater to keep an area of water clear of ice for gaseous exchange.

Therefore, even if you choose your fish carefully from a reputable dealer, there is still a risk that they may be carrying a low-level infection involving one or more potentially dangerous pathogens. Although often overlooked, this risk is a real one.

During the quarantine period, keep the new fish completely separate from all other fish for at least four weeks. Take care to prevent the transfer of disease organisms on contaminated equipment used to maintain the tank.. In fact, it is important to have a complete set of equipment, such as nets, bucket, scraper, siphon tube, etc., exclusively for use in the quarantine tank. Furthermore, the likelihood of transferring disease organisms from new fish to resident fish will be reduced if you carry out any routine maintenance on the quarantine tank after (*never* before) attending to the main aquarium or pond. Personal hygiene is also important, as fish disease organisms can be carried on wet hands. Always wash your hands before and after maintenance.

While they are in quarantine, observe the fish closely for any unusual symptoms and behaviour. It is much easier to deal with any signs of disease in the quarantine tank than in a set-up aquarium or pond.

Generally speaking, low temperatures slow down the life cycles of most fish pathogens, which means that any symptoms of disease will take longer to show at lower temperatures. Ideally, quarantine tropical fishes at 22-25°C(72-77°F) and coldwater fishes at no less than 12-15°C(54-59°F). At lower temperatures it is best to double the quoted four-week quarantine period.

If after four weeks (longer below 12°C/54°F), the quarantined fish have shown no signs of disease, carefully introduce them into the set-up aquarium or pond. Then thoroughly rinse the quarantine tank and *all* the associated equipment in running water and store dry for future use. Remember that even after quarantining the fish for four weeks (and perhaps treating them with one or more remedies) they are unlikely to be totally free of pathogens. Latent infections (caused by viruses, for example) may remain in or on their bodies, which highlights the importance of correct care in enhancing their natural resistance and thus preventing outbreaks of disease.

Aquarium fish

It is relatively simple to set up a quarantine tank for aquarium fish and small pond fish. All that is required is a small to medium-sized tank with a lid (45-90 litres/10-20 gallons is generally adequate), a foam cartridge filter run from an air pump or a small internal power filter, a heater-thermostat (for tropical fish), a thermometer and one or two plastic plants or half flowerpots to act as refuges for the fish. A fairly spartan tank will make cleaning and disinfection after quarantining each batch of fish much easier.

Controlling the build up of toxic fish wastes can be achieved by adding some zeolite granules (for freshwater aquariums only) to the filter chamber or suspended in a mesh bag. Alternatively, install a biological filter or filter medium that has been matured in the main tank containing fish. For marines, it is particularly important that the quarantine tank is a miniature

Basic quarantine or hospital tank

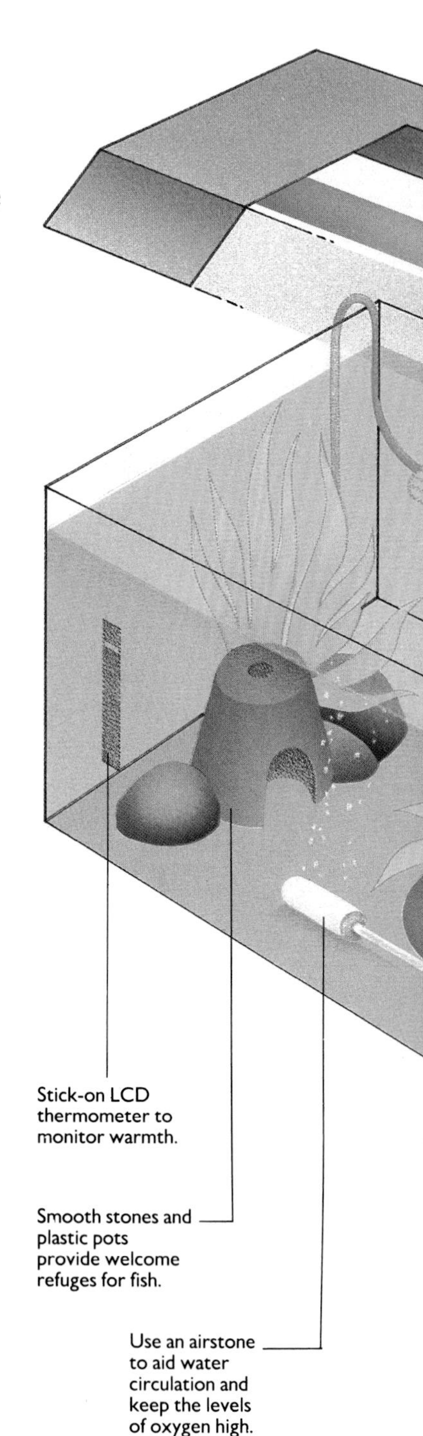

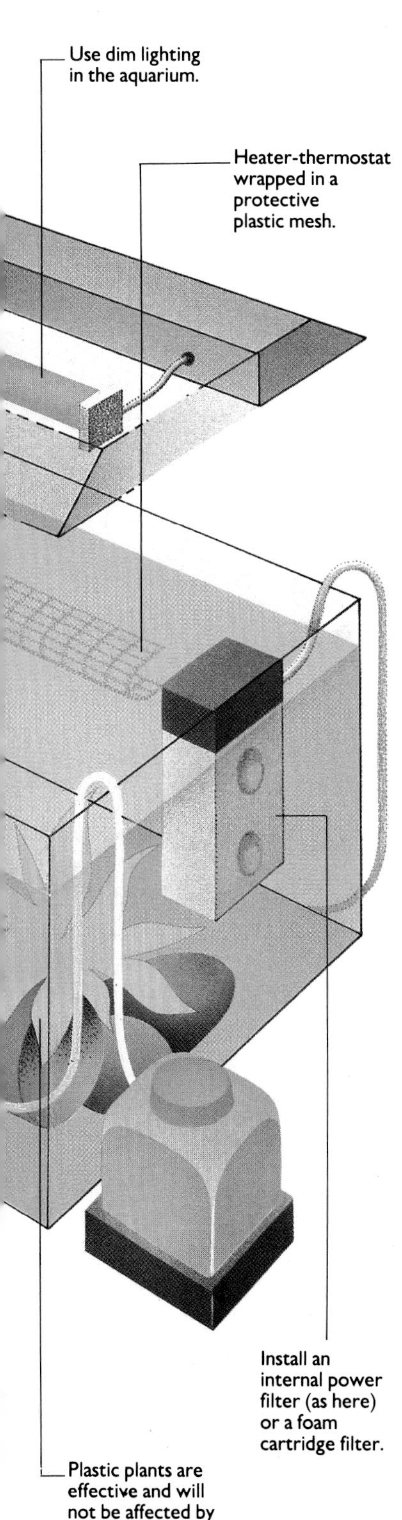

Pond fish

In order to quarantine all but the smallest pond fish, you will need larger facilities than those described above. A larger aquarium (at least 136 litres/30 gallons) will suffice in some instances, although other receptacles, such as a childrens' paddling pool or a large cardboard box lined with a pond liner, can be pressed into service at short notice. Whatever is used it is important to cover the container with a fine meshed nylon net, stretched tight and weighted down. This will keep the fish in, and the children, cats and birds out! Filtration is probably not required for most pond fish, although aeration is very important, particularly during warm weather. Regular partial water changes will, however, be necessary. If you are quarantining fishes in the garden, keep them out of the sun.

Plants

If they have come from an environment inhabited by fish, plants may also introduce pests and diseases. Ideally, therefore, wash new plants in lukewarm water and quarantine them in a fish-free tank at room temperature for several days. A more specific way of dealing with any pathogens present is to treat the plants with a plant disinfectant or mild broad-spectrum antiparasite remedy during the quarantine period. Alternatively, dip them in a weak (pale pink) solution of potassium permanganate for several minutes and then rinse them in running water. Another method is to soak the plants for an hour in a solution of potassium aluminium sulphate (alum) at 35°C(95°) and then rinse in clean water.

General hints on quarantine

- Avoid overstocking during quarantine, since this will only add to any disease problems which may occur.
- Locate the quarantine tank away from the set-up tank or pond where it will not be frequently disturbed.
- Maintain a watchful eye for any signs of overcrowding and oxygen shortage, such as fish gasping at the surface.
- Offer regular small meals throughout the quarantine period – using safe live foods to tempt their appetite, if necessary.
- Make twice-weekly partial water changes of 25-50 percent
- Adding new water at each partial water change will help to acclimate new freshwater tropical fish to your local water conditions. Tapwater conditioners will also help to acclimate freshwater fish.
- Since some conditioners reduce the effectiveness of certain disease treatments (check the manufacturer's instructions), it may be necessary to condition tapwater by aeration and/or by allowing it to stand at room temperature for a few hours. At least the chlorine will dissipate.
- If you are using a disease treatment, you may need to dose the new water in order to maintain the required concentration in the quarantine tank. Again, check the manufacturer's instructions.

CHAPTER 5

RECOGNIZING ILL-HEALTH

Making a prompt and accurate diagnosis is vital for successful treatment and long-term prevention of fish disease problems. While it is possible to make a post-mortem examination and/or send samples for detailed investigation (see pages 76-80 for further details), most fishkeepers need to be able to diagnose a problem from obvious external symptoms in live fish rather than resorting to such scientific procedures.

In this chapter, we consider first how to check fishes for health problems and how to categorize their symptoms in general. Then we consider how to submit material for post-mortem examination. For fishkeepers willing to make their own investigations we include guidance on using a simple light microscope. We also outline methods for anaesthetising and humanely killing fish

The remaining part of the chapter consists of a number of disease or problem recognition charts that provide useful 'jumping off points' to the detailed descriptions given in Chapter 6. For more details of how to use these charts, see the *How to use this book* section on pages 8-9.

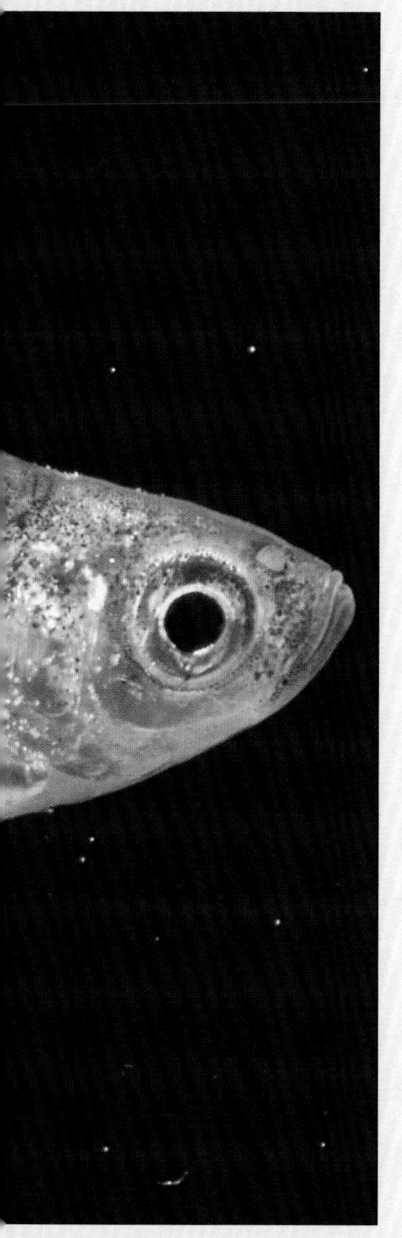

Acute or chronic?

Fish losses in an aquarium or pond often follow one of two broad patterns: acute or chronic.

Acute problems are apparent when all or most of the fish begin showing symptoms within a few hours to a day or so of the 'incident'. These symptoms will usually occur in a range of species and will, more often than not, be seen in the behaviour of the fishes rather than as external symptoms. All of these factors tend to suggest an environmental or water quality problem, such as high ammonia levels, oxygen shortage, poisoning from outside, etc.

Chronic problems manifest themselves over a longer period of time. Symptoms and/or losses may slowly increase over a number of days or even weeks and may be restricted to a single species or a group of related species, leaving other fishes in the aquarium or pond largely unaffected. This sequence of events would suggest the presence of some type of infectious disease. However, the disease itself could be caused by, or aggravated by, unsatisfactory environmental conditions. There are, however, some chronic disease problems, such as tumours, 'pop-eye' and nutritional disorders, that are either not infectious or have such a low infectivity that they may not follow this typical sequence of events. Usually such disorders affect only one or a few fishes at one time.

Examining fishes for symptoms

Recognizing a problem as either acute or chronic can be useful in the early stages of making a diagnosis, but it is also a good idea to establish a regular discipline of spending a few minutes each day just watching the fishes and noting any unusual behaviour or obvious symptoms of disease. This will enable you to spot the early stages of a problem and so apply the appropriate course of treatment as soon as possible. Use the naked eye, along with a x5-x10

Left: Some signs of ill-health are unmistakable. This rasbora shows the classic 'gold dust' of velvet disease; regrettably, not all health problems have such convenient 'signals' to alert the watchful fishkeeper.

magnifying glass, if available, to observe the fishes. Such vigilance can lead to prompt diagnosis and successful treatment. Feeding time provides a good opportunity to check your fishes' health.

Environmental influences

As we have seen in Chapters 3 and 4, poor or inadequate environmental conditions will predispose fish to ill-health, if not actually cause many fish disease problems. Therefore, details of past and present pond or aquarium conditions can be very useful in suggesting the likely cause of a problem. Such environmental influences encompass diet and feeding regime (including overfeeding), water temperature, stocking level, fish compatability, pH value, water hardness, ammonia and/or nitrite level, salinity (in marine aquariums), aeration, impurities in the water (copper, chlorine, etc.), any recent introductions of plants or fishes, and details of routine pond or aquarium maintenance. You will need to have information available on all these factors when seeking advice on the diagnosis of a particular disease. With this in mind, the questionnaire on page 81 indicates the kind of information which will help to produce an accurate diagnosis. Such information is often vital in eliminating the factors at the root of a problem.

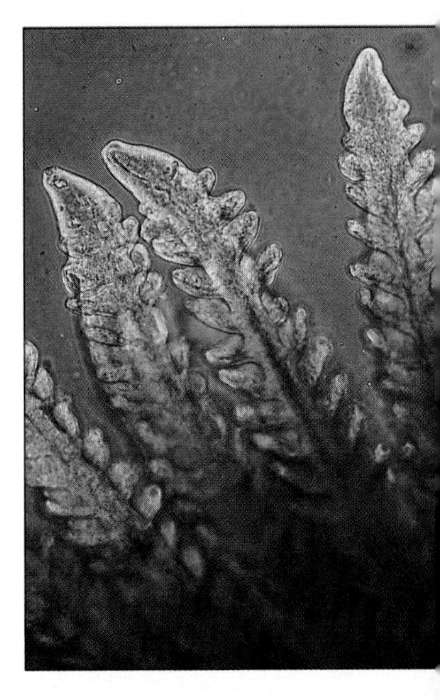

Above: Taking a closer look under a microscope can reveal vital evidence to clarify a diagnosis. This is what healthy gill tissue looks like at x25 magnification.

Making a diagnosis

For most hobbyists, making a disease diagnosis will include initial observations of symptoms or fish losses, a check on the prevailing environmental conditions (and a comparison with previous data), and an assessment as to whether the problem appears to be chronic or acute. This should then lead to a prompt, rational approach to treatment. Some fishkeepers, however, may also wish to send one or two specimens for laboratory examination.

Preparing material for laboratory examination

Check with your local veterinarian or fish health service before submitting live or dead fish for laboratory examination. Details of fish health services occasionally appear in aquarium and koi magazines. The quality of these private services may vary greatly, so try and get a recommendation.

Without doubt, live but obviously diseased fishes provide the best material for clinical and laboratory examination. However, for various reasons, this is not always possible – in which case seek professional advice regarding suitable methods for preserving dead material for examination. Some veterinarians and fish health laboratories can supply you with fixatives and other chemicals (e.g. viral and bacterial transport media). Deep freezing, by placing the dead fish (sealed in a polythene bag) within a domestic freezer, provides an alternative to chemical preservation. However this method of 'cryo-preservation' damages the fish's cells and tissues and may disrupt certain parasites, thereby limiting the extent to which the cryo-preserved fish can be used for post-mortem examination.

If a fish has been dead for 30 minutes or more its tissues will have significantly decomposed and this can complicate or render useless certain laboratory tests. Also, many skin and gill parasites will leave a fish within as little as a few minutes of its death.

Above: At a slightly higher magnification, another specimen of fresh gill tissue shows signs of haemorrhagic gill disease, with burst blood vessels (some highlighted with arrows) clearly visible in the individual lamellae.

The initial diagnosis

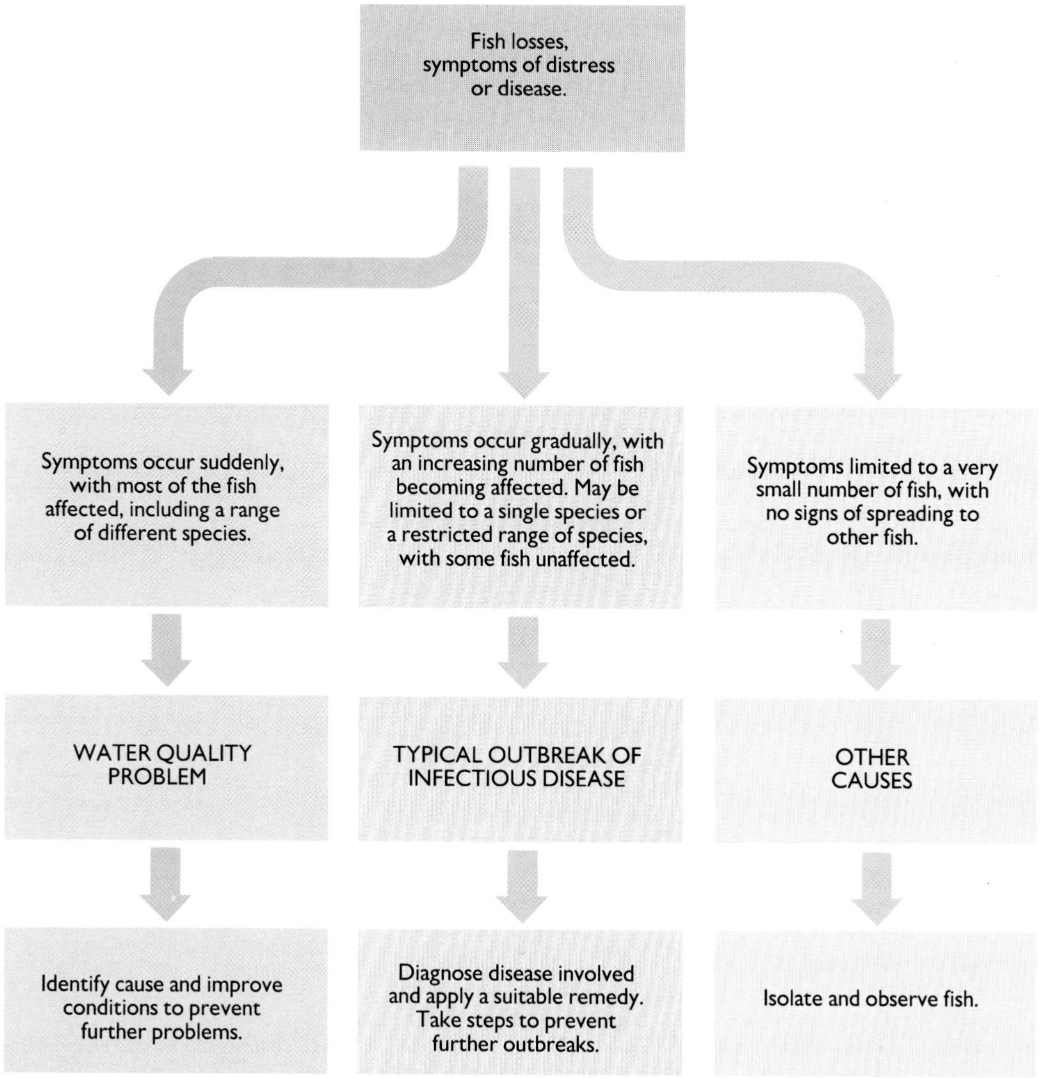

Fish losses,
symptoms of distress
or disease.

Symptoms occur suddenly, with most of the fish affected, including a range of different species.

WATER QUALITY PROBLEM

Identify cause and improve conditions to prevent further problems.

Symptoms occur gradually, with an increasing number of fish becoming affected. May be limited to a single species or a restricted range of species, with some fish unaffected.

TYPICAL OUTBREAK OF INFECTIOUS DISEASE

Diagnose disease involved and apply a suitable remedy. Take steps to prevent further outbreaks.

Symptoms limited to a very small number of fish, with no signs of spreading to other fish.

OTHER CAUSES

Isolate and observe fish.

Above: Rather than jumping to conclusions about the cause of a particular health problem, it is far better to consider the options in a rational way based on simple observations. This flowchart plots the logical links involved in making a broad initial diagnosis.

It often helps to submit a water sample from the aquarium or pond for laboratory analysis. The sample should be stored in a clean screw-capped glass container. About a cup-full of water is generally adequate.

Note: the water used to transport the fish is unsuitable for chemical analysis.

Humane methods of killing a fish

There seems little doubt that fish are capable of experiencing pain and stress. Hence, fish should be put out of their misery if they are badly injured or diseased such that recovery is highly unlikely. This 'life or death' decision can sometimes be a difficult one to make, and in such cases the hobbyist may wish to seek advice from an expert fishkeeper or veterinarian. Humane methods for killing fish are as follows:

- Concussion. The fish is restrained out of water by gently wrapping a sheet of wetted paper tissue around its body. Rest the fish on a firm surface and strike its head with a hard object such as a hammer. The aim is to instantly destroy the brain. Although it may seem barbaric, this method is swift and effective if performed correctly. Understandably, some hobbyists feel squeamish about performing this procedure in which case they should not attempt it.
- Overdose of anaesthetic. A suitable fish anaesthetic (see box) is administered at a sufficiently high concentration so as to cause loss of consciousness and eventual death. This method requires skill in selecting an appropriate concentration of anaesthetic and in assessing when the anaesthetised fish is actually dead. Hence, it should only be undertaken by a veterinarian or by someone who has received proper training in fish anaesthesia.

Some widely advocated methods of fish euthanasia are no longer considered humane. These include: slow freezing; dropping the fish into very hot or ice-cold water; breaking its backbone; decapitation. All of these aforementioned methods are believed to cause suffering to the fish. It is equally cruel to leave a fish out of water to die or to flush it alive down the toilet.

Despatching preserved material

Having contacted the laboratory and preserved the material for examination, you should then consider how to despatch it. Usually, it is possible to send preserved material – either frozen or fixed – in the mail, although check this first. It should be possible to send frozen material inside two sealed plastic bags in a well-insulated, leakproof container. Preparing chemically fixed specimens for despatch takes a little more care. Once the material has been in the fixative for at least 48 hours, you can remove it and safely dispose of the liquid. Wrap each preserved specimen in a soft cloth soaked in fixative and seal it inside two plastic bags and send it off in a leakproof container.

Whenever you send fishes for post-mortem examination, it is vital that you supply all available information on the history of the specimen, symptoms of disease, recent pond or aquarium care, etc.

Safe disposal of dead fishes

Never flush dead or dying fishes down the toilet, release them into natural waters or simply discard them in the waste bin. This type of disposal may lead to novel diseases being introduced to native fish populations, perhaps with disastrous results. If possible, dispose of dead fishes by incineration. Alternatively, wrap them in newspaper generously sprinkled with household disinfectant or bleach (wear protective gloves) and seal the package in two plastic bags. This should be quite safe to put in a household waste bin, but be sure to secure the lid to prevent contact with dogs, cats and vermin before the waste is collected.

Since some fish disease organisms can infect humans, be sure to handle discarded fish extremely carefully. In fact, always take sensible hygiene precautions when working with fish. Cover cuts and abrasions on your hands and arms, for example, and wash thoroughly after carrying out routine maintenance.

Anaesthetics and destruction of fish

Chemicals such as ethyl-m-aminobenzoate, benzocaine and quinaldine can be used as anaesthetics for fish, thus permitting surgery, aiding invasive treatment and sampling methods, and reducing the trauma of transport and handling. However, the activity of each anaesthetic often depends on a number of factors, particularly the fish species to be anaesthetized and the temperature of the water. As a result, it is difficult to provide any general guidelines on the necessary procedures. Hobbyists wishing to anaesthetize their fish should therefore contact a veterinarian, or someone familiar with fish anaesthesia, before undertaking any trials.

From time to time, it may be necessary to kill a fish, perhaps for submission for post-mortem examination or perhaps when further treatment seems pointless. An overdose of one of the above anaesthetics is a painless method of destruction, and a veterinarian should be able to advise and/or assist you. Such anaesthetics added to a small volume of water containing the fish cause the fish to rapidly lose equilibrium, the gill movements cease and death soon follows.

If a suitable anaesthetic is not available, you can quickly destroy small fish by decapitation with a sharp knife or scissors. Stun a large fish first with a sharp blow to the head using a wooden or metal rod, and then decapitate it. As mentioned in Chapter 5, page 78, always dispose of unwanted fish (whether dead or alive) in a sensible fashion.

Below: Anaesthetizing a koi in a poolside container in order to manually remove anchor worms. If in any doubt, seek expert help.

Post-mortem techniques

Performing a post-mortem on a fish requires specialist skill and hence is best left to a veterinary surgeon or suitably qualified fish health professional. A post-mortem typically involves external and internal examination of the fish, sometimes augmented by bacteriological or virological sampling of tissues, where appropriate. In certain situations, small pieces of tissue or organ are removed for histology, a process that involves microscopical examination of chemically stained, ultra-thin sections of tissues.

Fish anaesthesia

Certain chemical anaesthetics are suitable for use on fish, thus permitting surgery or aiding invasive treatments and sampling methods. In some cases, mild anaesthesia is useful for reducing the trauma of transporting and handling fish.

Four commonly used fish anaesthetics are:

- Benzocaine. In its pure form this is a white powder. It is poorly soluble in water so must initially be prepared as a stock solution in alcohol.

- Tricaine methane sulphonate (TMS). Also known as MS222. This is also a white powder. Although more expensive than benzocaine this chemical is readily soluble in water.

- 2-Phenoxy-ethanol (2-PE). Also known as phenoxythol. A viscous liquid that is moderately soluble in water.

- Eugenol (clove oil). This oily chemical has the advantage of being available without prescription. It is sold for anaesthetising koi, but can be used on other fish. It is also widely available (as clove oil) from pharmacies. The minimum recommended dose for clove oil is 10 drops per litre of water (45 drops per imperial gallon). The oil does not dissolve readily so must be thoroughly mixed in a small volume of warm water before adding to the anaesthetic container.

The availability of fish anaesthetics will vary according to country, so check with your veterinarian or aquarium store. If you are preparing the fish for surgery or some other potentially painful procedure then it is vital to check that the anaesthetic has analgesic properties (i.e. it is known to suppress pain in fish). The three anaesthetics listed above all have analgesic function.

The correct dosage will depend on the type of chemical anaesthetic used, and is also influenced by factors such as the species of fish and water temperature. As a result it is difficult to provide any general guidelines.

The fish must be closely monitored during anaesthesia until it reaches the desired level of anaesthesia. Assessing the depth of anaesthesia in fish is an acquired skill that is best left to a veterinarian or other suitably qualified professional. Fish may suffer unnecessary pain or even death if anaesthesia is not carried out properly.

An anaesthetic can be administered at an 'overdose' level in order to humanely kill a fish. Again, this method of euthanasia should only be undertaken by someone with appropriate training.

Right: Terminally ill molly. Such fish should be put out of their misery using an approved method of euthanasia, such as an overdose of anaesthetic.

A trouble-shooting questionnaire

The following questionnaire is a basic template that you can adapt to your own special situation. It is designed to be photocopied so you can append the appropriate information on your photocopy. The answers to the questions should enable an experienced fishkeeper or fish health professional to identify likely cause(s) of the problem or at least to eliminate the most obvious causes. It usually helps if you can also provide one or two live specimens of sick fish for examination. Of course, the questionnaire is not fool-proof and it is vital to view each problem in the light of the general points made in the diagnosis chart on page 77.

Background information

1 How many years of fishkeeping experience do you have?.............

2 Local tapwater conditions (if known):
You can measure these parameters using simple aquarium test kits:
pH value:..................................
hardness (state hardness scale used):..
nitrate level:..............................

3 Do you always condition new tapwater (i.e. to remove chlorine/chloramine) before adding to the aquarium or pond?
Yes/No..

Current problem

4 Type of system affected, eg:
pond; ....
coldwater aquarium;.....................
tropical freshwater aquarium;.....
tropical marine aquarium ....

5 When did the problem begin?
....
....

6 List the symptoms seen.
For example: list any abnormal lumps, ulcers, markings or damage on the fish's skin or fins; any unusual behaviour or unusual swimming posture (e.g. fish gasps at surface, etc.).
State if the affected fish are still feeding. Also state if any fish have already died of the problem.......................................
....
....
....
....
....
....
....
....

7 If several fish are affected, did they all develop the problem suddenly (e.g. within a 48 hour period) or over several days or weeks? ....
Yes/No ....

8 Which are the worst affected species?.......................................
....
....
....
....

9 List any unaffected species in the same aquarium or pond.
....
....
....
....

10 Are any of your other aquariums or ponds affected by the problem? ....
....
....

Conditions in the affected tank or pond

11 Size of aquarium/pond
State size or volume (gallons or litres).......................................

12 How long has the aquarium or pond been set up? ....
....
....

13 Water conditions in the aquarium or pond (if known):
pH value:.....................................
Hardness:....................................
Ammonia (state date of test):
....
Nitrite (state date of test):
....
Nitrate (state date of test):
....
Temperature:..............................

14 Stocking level (number and types of fish): ....
....
....
....
....
....
....
....
....

15 Is biological filtration installed?
Yes/No ....
If yes, what type of biological filter (e.g. undergravel filter, air-driven sponge filter, motorized canister filter, etc.)
....
....
When was the filter medium last cleaned? ....
....

16 Is the aquarium or pond aerated?
Yes/No ....

17 Do you perform regular partial water changes? ....
If yes, how much and how often?
....
....
....

18 When did you last add new fish to the aquarium/pond?
....
....
....

19 Have you added live plants recently (within four weeks of current problem)? ....
....
....
....

20 Do you feed live foods? (e.g. tubifex worms, daphnia)
If so, which types:
....
....

21 Could any of the following household or garden chemicals have gained access to the aquarium or pond:
Aerosols or sprays (especially insecticides or herbicides)
Paint, glue or varnish fumes
Garden run-off
Other fumes or chemicals
....
....

22 Have you added any décor (rocks, sand, ornaments) recently?
Yes/No ....
If so, list any items of décor NOT sold for aquarium/pond use? ....
....
....
....

23 Have you added disease remedies recently (e.g. within four weeks of the current problem)?
List any remedies used with dates of application: ....
....
....
....
....

24 In the case of aquariums, does the affected tank house ornamental invertebrates (e.g. freshwater or marine crabs or prawns; corals, etc.).
Such information may be important should chemical disease treatments be necessary. Yes/No ....

25 Give any other comments or observations that may be relevant to the current problem:.................................
....
....
....
....
....

Using a compound light microscope

A compound light microscope is so-called because the magnification is a compound of the lens power in the eyepiece and the objective, and also because light is reflected through the preparation from below. It is a valuable aid to investigating the cause of ill-health in fish, being able to provide magnifications in the range of ×25 to ×1000. It is best to start with the lowest power objective in order to get your bearings and then proceed carefully at higher powers. To guard against bringing the lens down onto the slide and squashing it, focus by moving the slide away from the lens, which may involve moving the tube or the stage, depending on design.

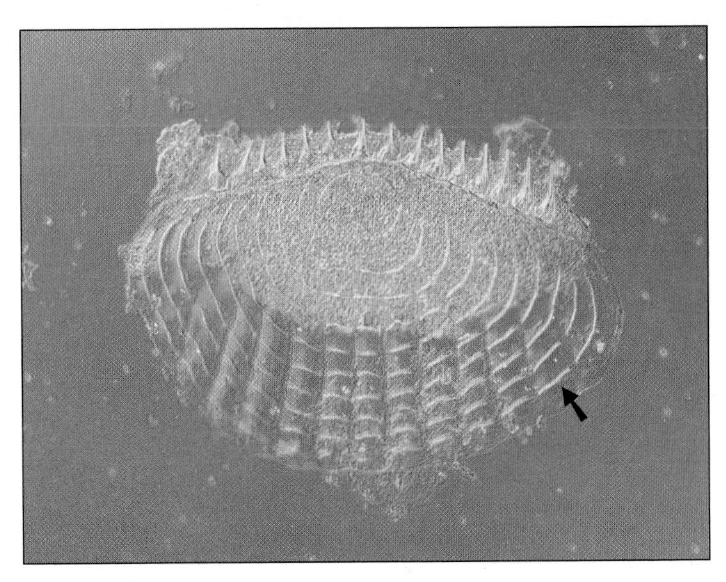

Above: A detached fish scale viewed under low power of the microscope. The concentric 'growth rings', which may be used to age the fish, can be clearly seen.

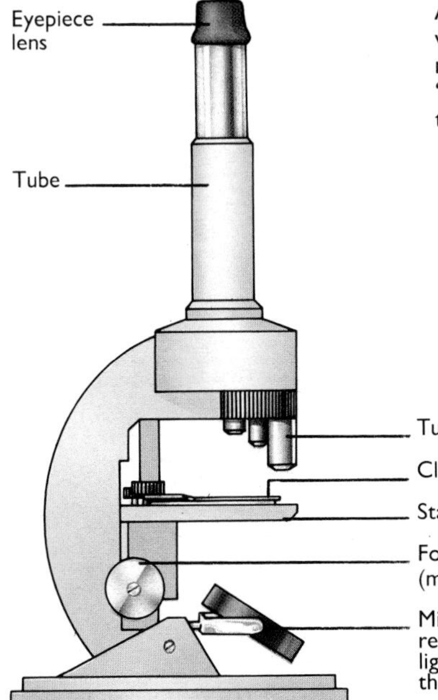

Right: A fresh skin smear at ×100 magnification. Note the fish scale at top right, the algal filament and many air bubbles.

SKIN AND FINS I

OBVIOUS PARASITES

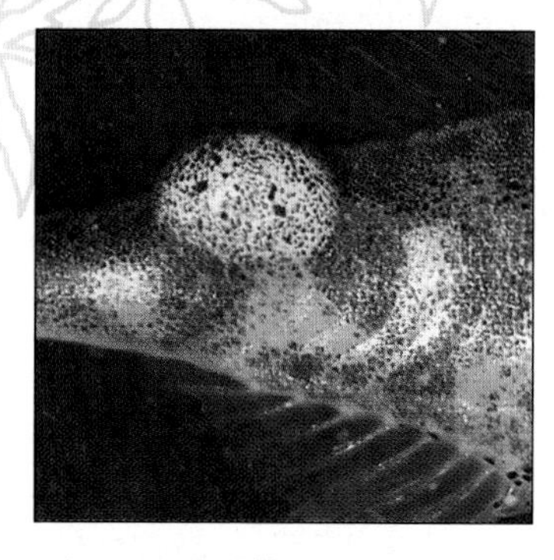

Spherical or oval, smooth yellowish cysts, up to 1cm(0.4in) across, on skin, fins and gills.
● **Nodular diseases** *page 128*

Dusting of gold speckles on skin and fins. Skin peeling away in strips.
● **Velvet disease** *page 158*
○ See also **White spot** *page 166*

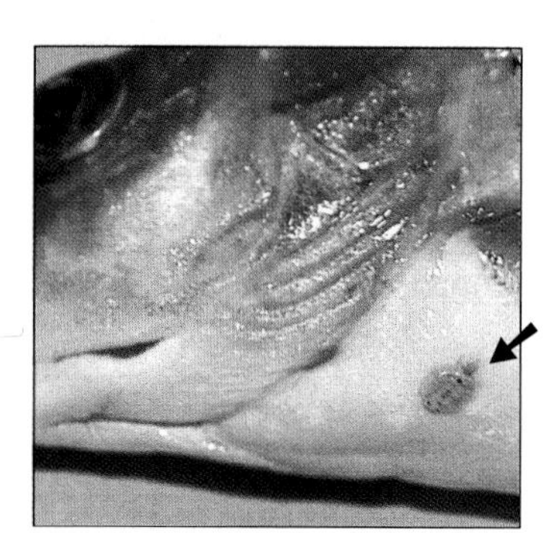

Disc-shaped parasite, up to 1cm (0.4in) across, clinging tightly to skin and fins. Reddened lesions where parasites have fed.

● **Fish lice** *page 112*

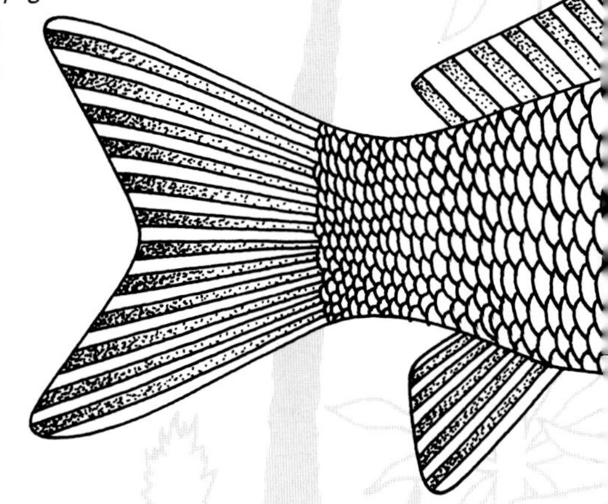

Sprinkling of white spots, each up to 1mm (0.04in) across, on skin, fins and gills.

● **White spot disease** *page 166*
○ See also **Guppy disease** *page 169*

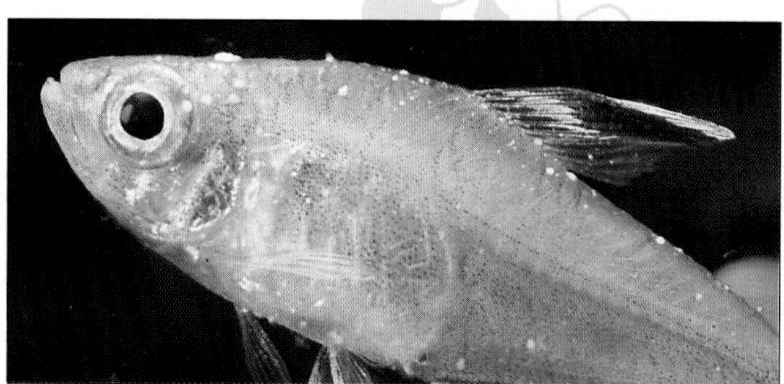

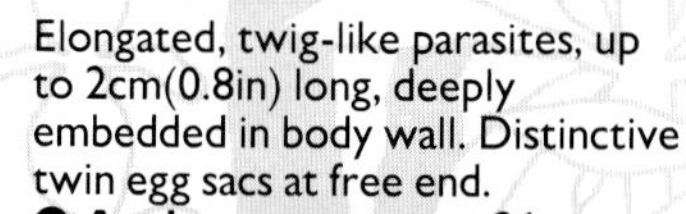

Elongated, twig-like parasites, up to 2cm(0.8in) long, deeply embedded in body wall. Distinctive twin egg sacs at free end.
● **Anchor worm** *page 96*

Worm-like parasites, up to 4cm (2in) long, attached to skin and fins by sucking discs at each end. Reddened areas indicate previous attachment points. Generally in ponds only.

● **Leech infestation** *page 124*

Fungus-like growths, especially around the mouth. Reddened ulcers on the body and frayed fins.

● **Cotton-wool disease** *page 100*

○ See also **Fish fungus** *page 108*

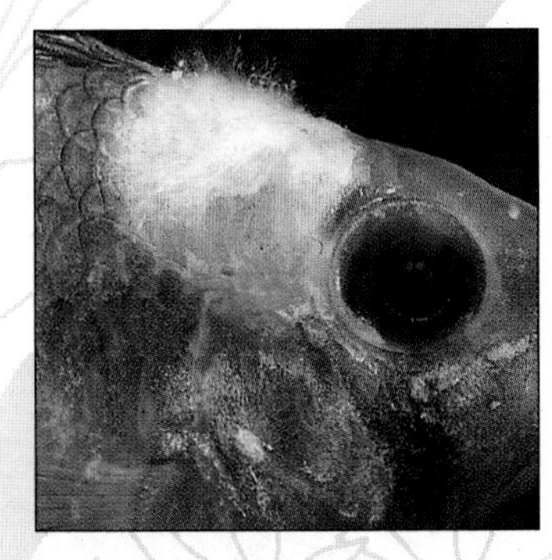

Cotton-wool-like tufts on the skin, usually white but possibly grey or brown in colour.

● **Fish fungus** *page 108*

○ See also **Cotton-wool disease** *page 100*

White, maggot-like parasites, up to several mm long, on gills, gill cover and inside mouth.

● **Gill maggots** *page 112*

Black spots (cysts), up to 2mm(0.08in) across, on skin and fins. Yellowish cysts also occur.

● **Black spot, etc.** *page 178*

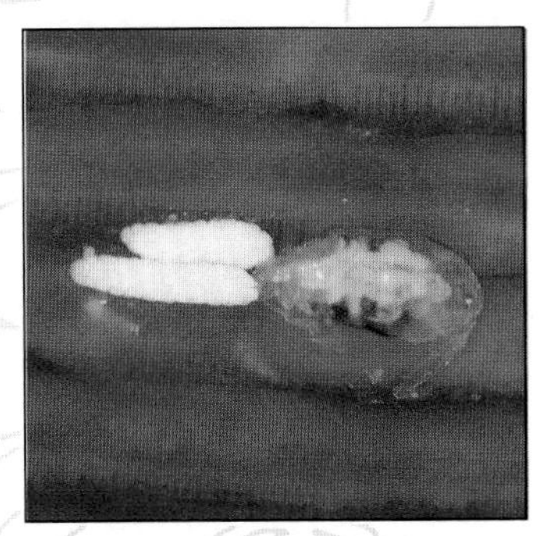

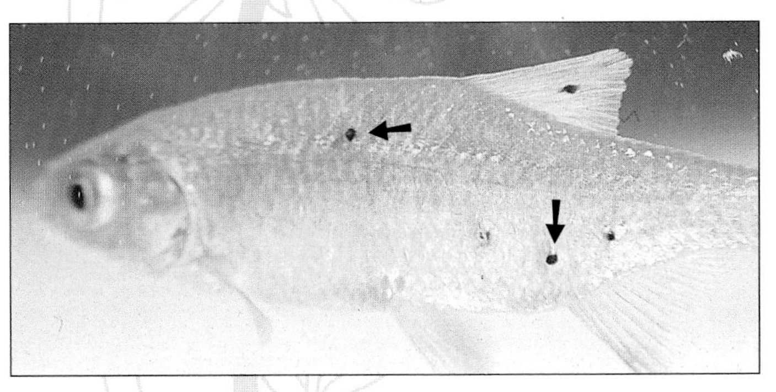

SKIN AND FINS 2

LESIONS, LUMPS AND BUMPS

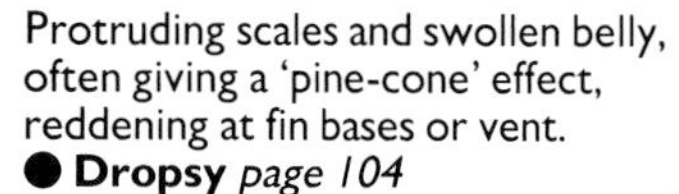

Protruding scales and swollen belly, often giving a 'pine-cone' effect, reddening at fin bases or vent.
● **Dropsy** *page 104*

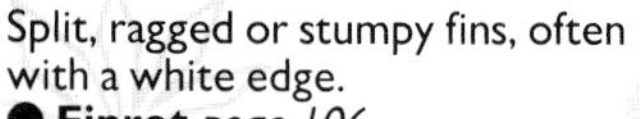

Split, ragged or stumpy fins, often with a white edge.
● **Finrot** *page 106*
○ See also **Cotton-wool disease** *page 100*

Pale shallow lesions on skin, plus other symptoms such as hollow-bellied appearance, 'pop-eye', colour loss and listlessness.
● **Wasting disease/Fish TB** *page 162*

Unusual growths or swellings clearly visible on any part of the fish's body.
● **Tumours** *page 152*

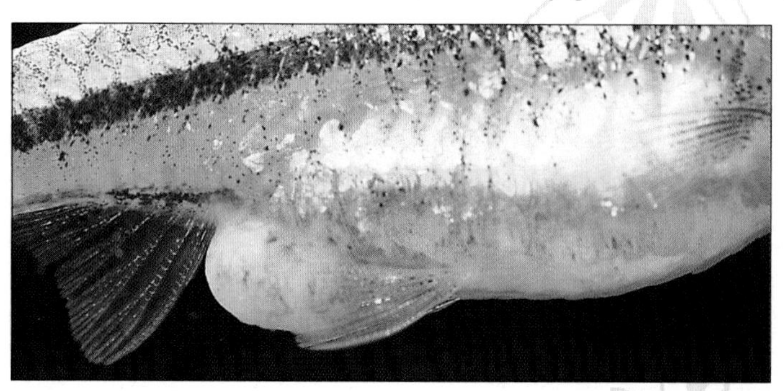

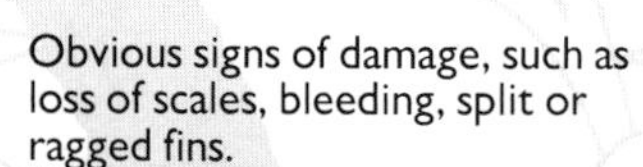

Obvious signs of damage, such as loss of scales, bleeding, split or ragged fins.
● **Physical damage** *page 142*

Grey-white film of excess mucus on skin, plus scratching and/or rapid gill movements.
● **Sliminesss of the skin** *page 146*

Small holes in the body, particularly in the head region. Lesions may enlarge and produce yellowish mucus trails.
● **Hole-in-the-head disease** *page 122*

Grey edges to fins and/or gills with no other symptoms. Only in fish that have been kept with freshwater mussels.
● **Glochidial infestation** *page 120*

Rough raspberry- or cauliflower-like growths on skin and fins.
● **Lymphocystis** *page 152*
○ See also **Fish pox** *page 116*

Smooth white, grey or pink growths on skin and fins often looking like molten wax. Extreme growths may take the colour of surrounding tissue.
● **Fish pox** *page 116*
○ See also **Lymphocystis** *page 152*

Lesions, ulcers or sores on the body, plus reddening at fin bases and vent.
● **Ulcer disease** *page 154*
○ See also **Physical damage** *page 142*

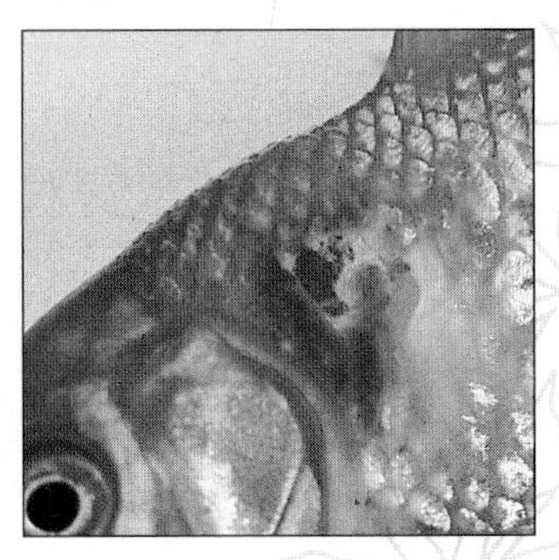

EYES

Bulging eye or eyes, together with swollen belly and raised scales.
● **Dropsy** *page 104*

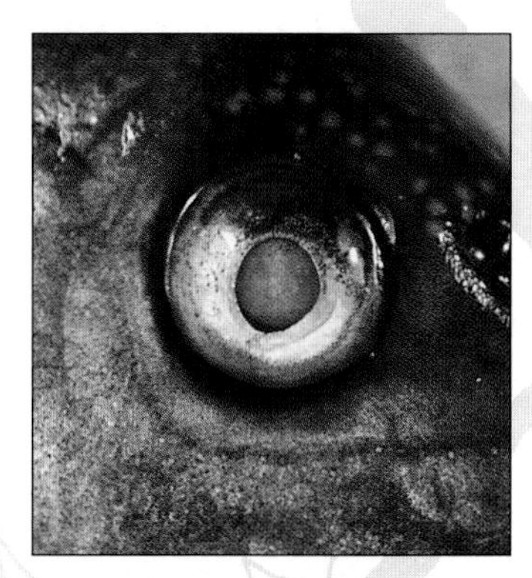

Cloudy lens, with no other symptoms.
● **Eye fluke** *page 178*

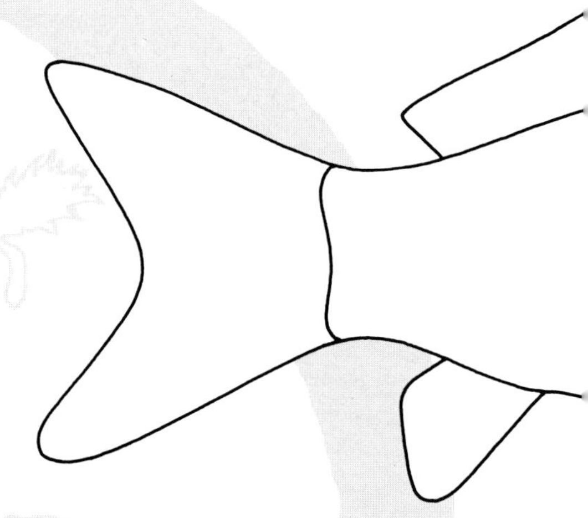

Eye lost from socket, with no other obvious symptoms.
● **Physical damage** *page 142*

One or both eyes protrude in an abnormal way, but no other obvious symptoms.
● **Pop-eye** *page 144*

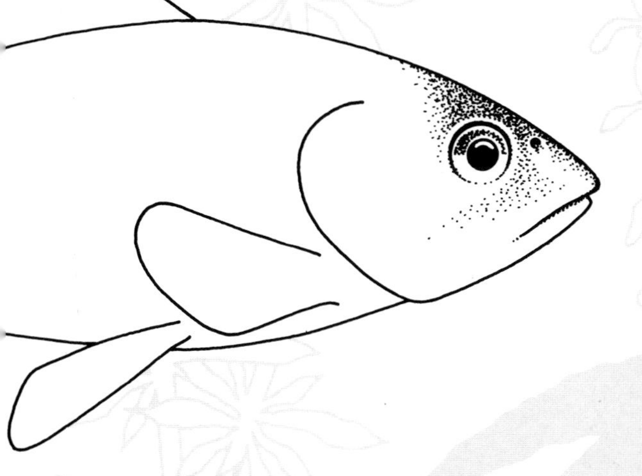

Bulging eye or eyes, together with skin lesions and emaciation.
● **Wasting disease/Fish TB** *page 162*

Cloudy eyes, together with poor growth and bleeding at fin bases.
● **Nutritional problems** *page 130*

SHAPE, COLOUR AND BEHAVIOUR

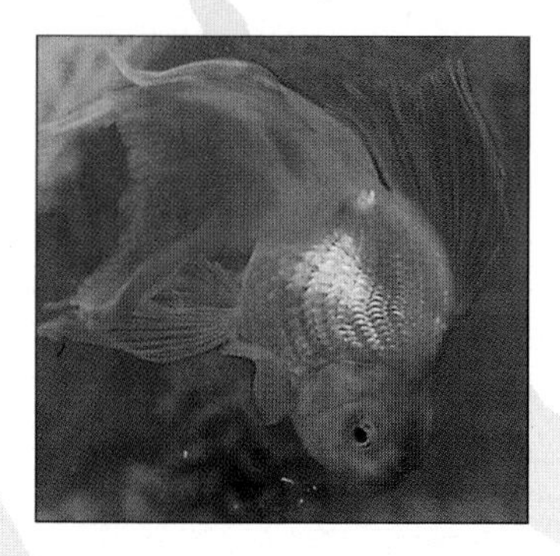

Unnatural swimming behaviour, 'listing' to one side or even floating on side or back.
● **Swimbladder disorders** *page 150*

Long, pale, faecal cast, perhaps with darkening of colour, pop-eye, and loss of appetite.
● **Dropsy** *page 104*
● **Hole-in-the-head disease** *page 122*

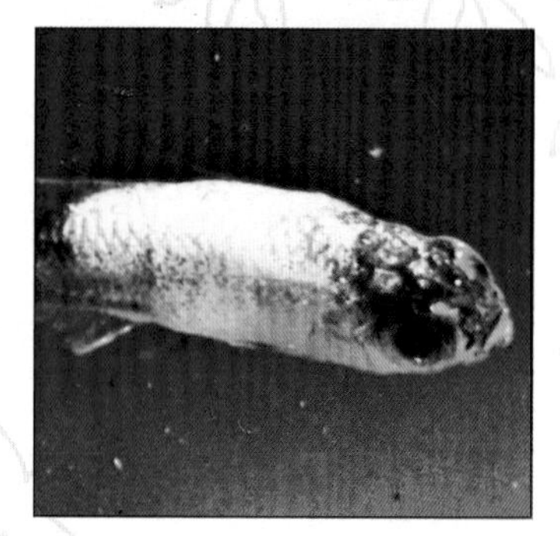

Loss of colour, unusual swimming behaviour, emaciation, spinal curvature and finrot.
● **Neon tetra disease** *page 126*

Swollen belly and impaired swimming behaviour. Scales do not stick out.
● **Worms in the body cavity** *page 170*

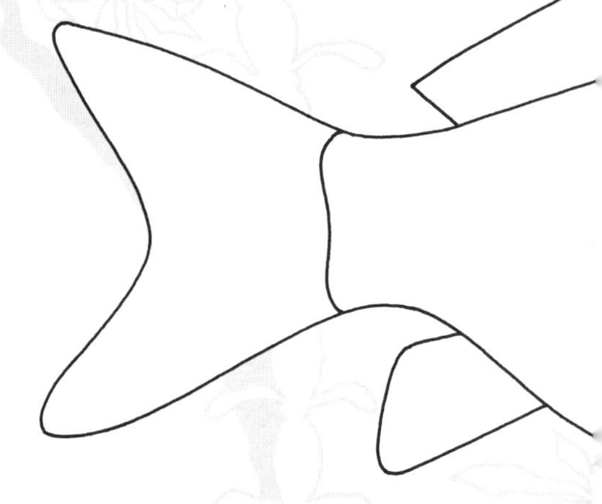

Peculiar swimming behaviour, rapid gill movements, gasping at water surface, inactive on bottom, generally 'off colour'.
● **Water quality problems** *page 164*

Red 'paintbrush' of worms protruding from vent, perhaps associated with emaciated appearance.
● **Worms in the intestine (Camallanus)** *page 172*

Thin, emaciated appearance without other obvious symptoms.
● **Worms in the intestine** *page 172*

Thin appearance, gaping gills, gasping at surface, listless behaviour.
● **Water quality problems** *page 164*
● **Gill disease** *page 118*

Bulges apparent within body (and also perhaps on the outside).
● **Nodular diseases** *page 128*

Thin, emaciated appearance, listless behaviour, anaemia and perhaps bulging eyes.
● **Blood parasites** *page 98*
● **Wasting disease/Fish TB** *page 162*
○ See also **Hole-in-the-head disease** *page 122*

Firm noticeable swellings that distort the body shape.
● **Tumours** *page 152*

Bloated body, darkening of coloration and lack of appetite.
● **Dropsy** *page 104*

INTERNAL ORGANS AND EGGS

Grey, brown or white woolly tufts on eggs, spreading from opaque dead eggs to affect live ones.
● **Egg fungus** *page 108*

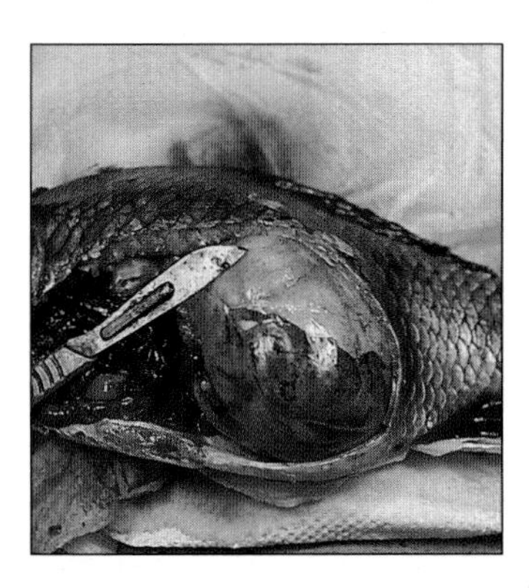

Firm growths among the internal organs in any part of the body.
● **Tumours** *page 152*

Ribbon-like or round worms visible within the intestinal tract.
● **Worms in the intestine** *page 172*

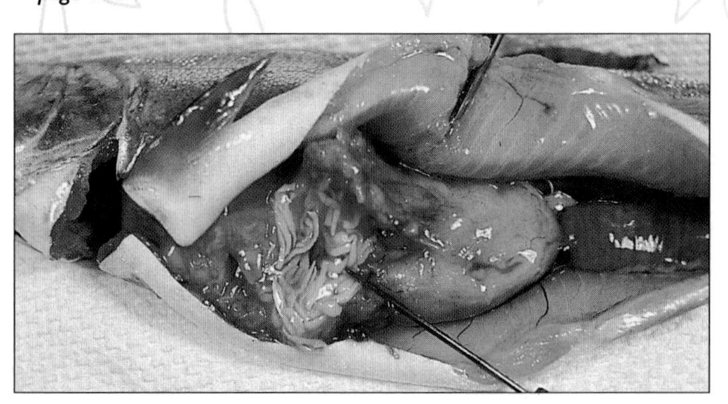

Whitish nodules within the internal organs of the body.
● **Wasting disease/Fish TB** *page 162*

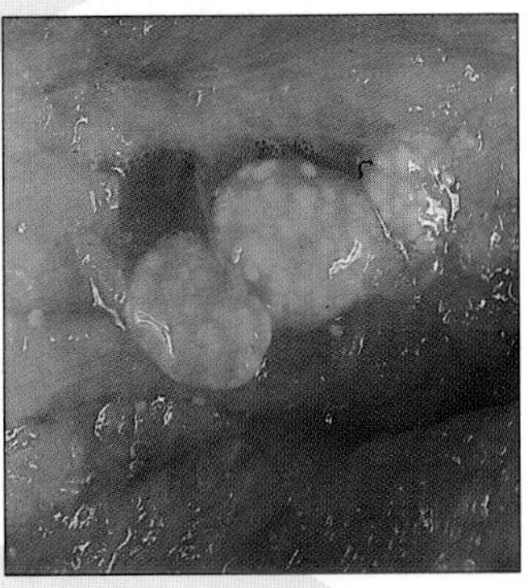

Yellowish white cysts up to 1cm(0.4in) across among internal organs of the body.
● **Nodular diseases** *page 128*

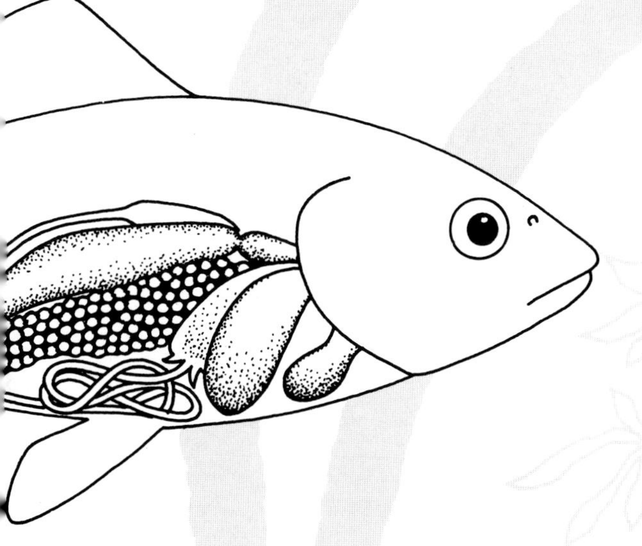

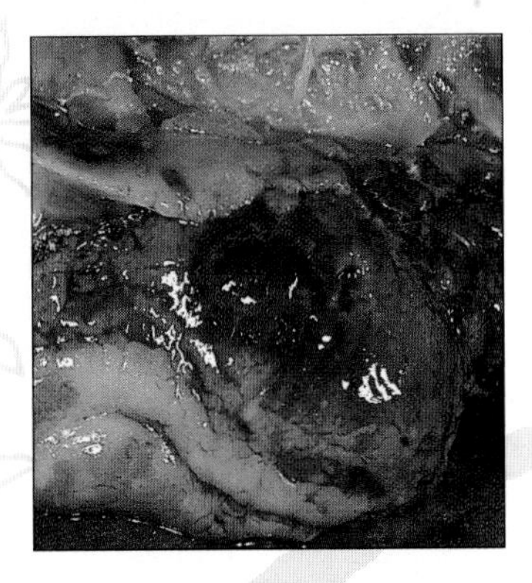

Bleeding among the internal organs and accumulation of fluid within the body cavity.
● **Haemorrhagic septicaemia** *page 154*

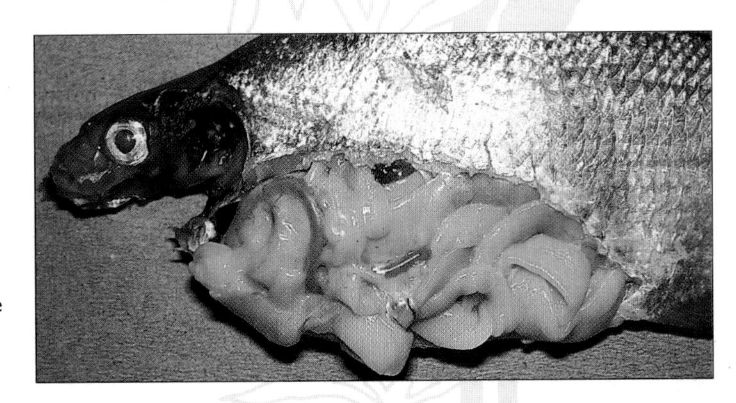

Ribbon-like or round worms in the body cavity of the fish.
● **Worms in the body cavity** *page 170*

PESTS

Green water, green threadlike growths, brownish filmlike growths and/or dark green slimy sheets over rocks, plants and substrate.
● **Algal problems** *page 132*

White, cream or orange flattened worms with arrow-shaped heads. Up to 1cm(0.4in) long.
● **Flatworms** *page 136*

Hordes of tiny, insect-like creatures about 1mm(0.04in) long clustered on the damp glass just above the waterline.
● **Mites** *page 136*

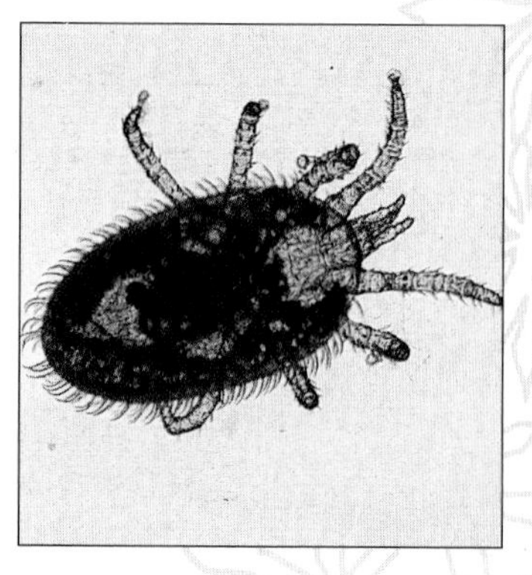

Soft-bodied, stalk-like polyps, each with a ring of tentacles, attached to plants and rocks. Up to about 2.5cm(1in) in height.
● **Hydra** *page 138*

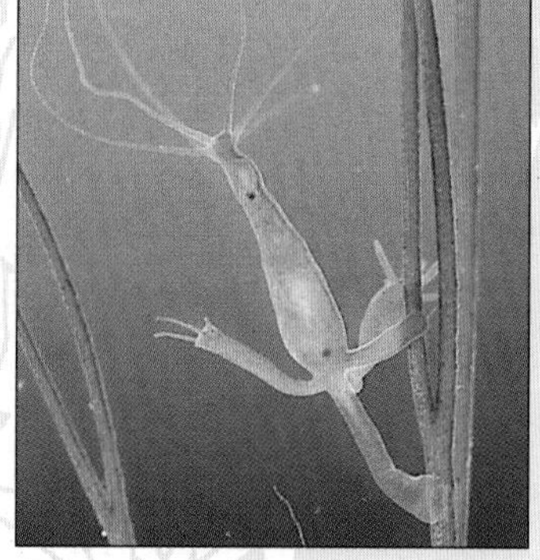

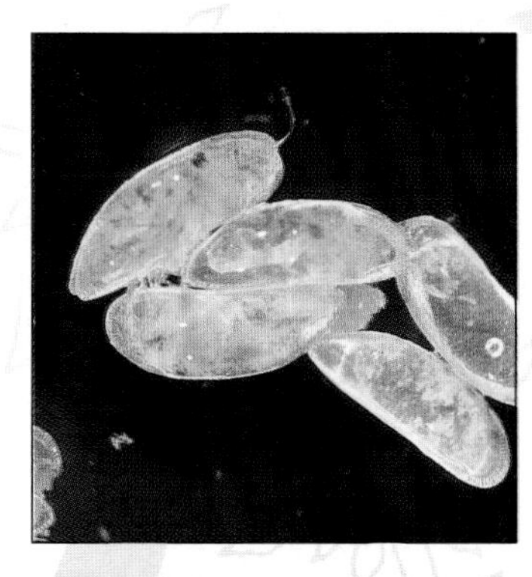

Bean-shaped creatures up to 3mm (0.1in) size. Scuttle over the substrate and plants.
● **Ostracods** *page 136*

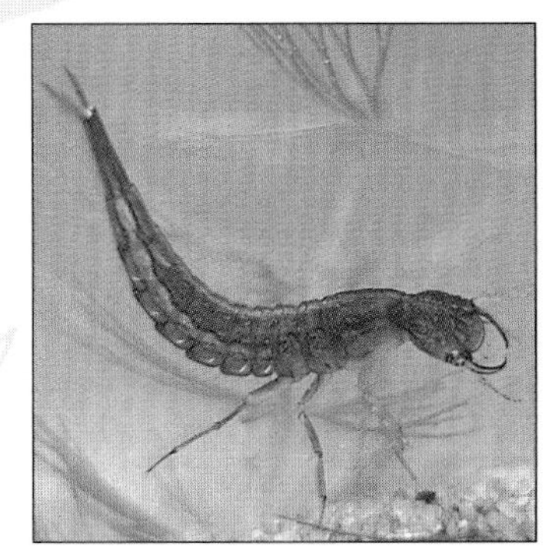

Larval beetles, such as *Dytiscus marginalis*, with fearsome-looking curved jaws. Up to 5cm(2in) long.
● **Beetles** *page 140*

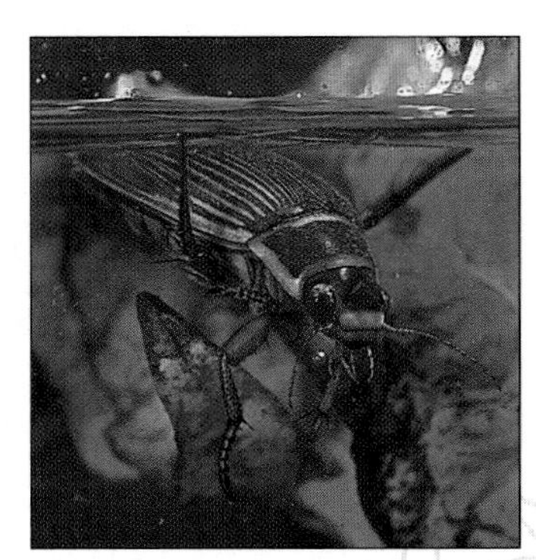

Adult beetles, such as *Dytiscus marginalis*, up to 3cm(1.2in) long.
● **Beetles** *page 140*

Dragonfly nymphs with retractable jaws, such as *Aeshna*, up to about 4cm(1.6in) long.
● **Dragonflies** *page 140*

Masses of whitish, coral-like polyps profusely ringed with tentacles. Up to several cm in height. Only in marine tanks.
● **Aiptasia** *page 138*

Large numbers of snails. Species such as the Malayan livebearing snail (*Melanoides*) may build up rapidly in numbers.
● **Snails** *page 140*

CHAPTER 6

AN A-Z OF COMMON PESTS AND DISEASES

In Chapter 5, we considered how to recognize the symptoms of the common problems that may affect pond or aquarium fish. In this chapter, we take the story one step further by looking at a wide range of pests and disease in sufficient detail for you to be able to confirm or reject your initial diagnosis and to consider what treatment or control measures to take. The pests and diseases are not presented in any scientific or taxonomic sequence; simply in A-Z order of the most widely used common names that best describe the physical appearance of the causative agent – anchor worm, for example – or the symptoms they cause, such as sliminess of the skin. The text for each condition is set out under four headings:

Caused by A simple statement of the most likely causative agent or agents. Some conditions are quite clearly caused by one particular organism – such as one species of fungus, bacterium, virus or parasite – whereas other health problems can arise from the action of one or several agents from a range of disease organisms or the combination of physical, developmental and environmental influences with or without the added complication of associated disease organisms.

Obvious symptoms/signs The signs of a health problem that can be clearly seen with the naked eye, including the appearance of visible parasites and any behavioural effects the condition causes.

Occurrence of the disease/problem A brief explanation of how and why a particular disease or problem occurs. If the condition is caused by a parasite, we include details of its life cycle and, in most cases, show this cycle in an accompanying illustration.

Treatment and control An outline of the most appropriate treatment or control measures. The treatment advice reflects ways of curing affected fishes – by carefully removing parasites or adding a suitable remedy to the water, for example – whereas control measures include practical ways of preventing further outbreaks – such as eliminating intermediate hosts in a parasitic life cycle, for example. Having noted the advice given in this section, it is vital that you also refer to Chapter 7 for further details of treatment and control methods before taking any action.

Left: This fancy goldfish is displaying signs of mouth fungus and finrot. In such situations, choose a broad-spectrum treatment or tackle the most pronounced symptoms first.

ANCHOR WORM

Caused by
The crustacean parasite *Lernaea*.

Obvious symptoms
Anchor worms are elongated parasites with two egg sacs at the posterior end. They usually occur embedded in the muscle of the body wall, often penetrating as far as the internal organs. A raised ulcer usually develops at the point of attachment and secondary infections may also occur here. Heavy infestations may cause weight loss and even death.

Occurrence of the disease
This parasite most often occurs on newly imported fish, and is more of a problem in garden ponds during the summer months than in aquariums.

Male *Lernaea* have a relatively short lifespan and die after mating. It is the longer lived female which is most often seen attached to the fish host, with her two pronounced egg sacs. The eggs hatch to release free-living juvenile parasites, which eventually moult to give rise to the adult stages. The juvenile stages can live without a fish host for at least five days, and *Lernaea* may overwinter as a female parasite on the fish or as eggs.

Treatment and control
Effective control is best achieved by using an organophosphorus insecticide, such as metriphonate (see Chapter 7), to eliminate the free-living juvenile stages and manually removing any attached adult *Lernaea*. Therefore, add a suitable remedy to the pond or aquarium, and remove and treat any fish with attached adult parasites. Carefully and swiftly remove each *Lernaea* with a pair of fine forceps. Grip the parasite firmly near to its point of attachment to the fish's body and pull it away. Dab the surrounding area with mercurochrome or a similar topical antiseptic. This manual removal

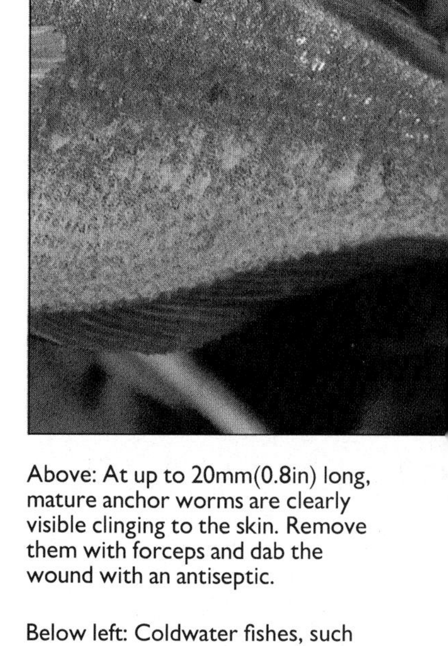

Above: At up to 20mm(0.8in) long, mature anchor worms are clearly visible clinging to the skin. Remove them with forceps and dab the wound with an antiseptic.

Below left: Coldwater fishes, such as this fancy goldfish, are more likely to be infested with anchor worms than are tropical fishes, usually when newly imported.

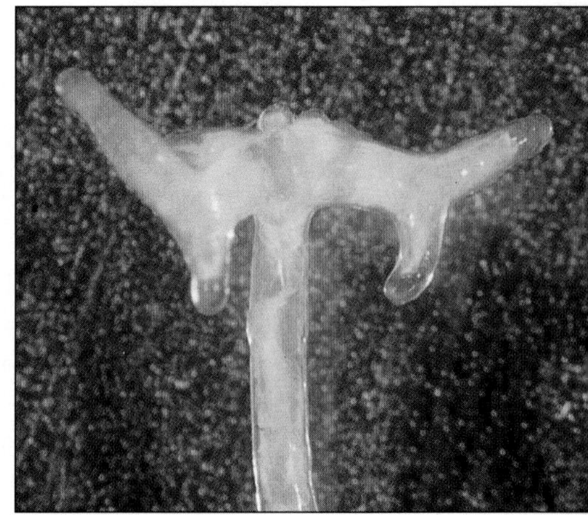

Below right: Close-up of the attachment organ of *Lernaea*.

Right: Heavy infestations of anchor worm can severely debilitate the host fish and make it more vulnerable to secondary bacterial and fungal infections. Be sure to start treatment early.

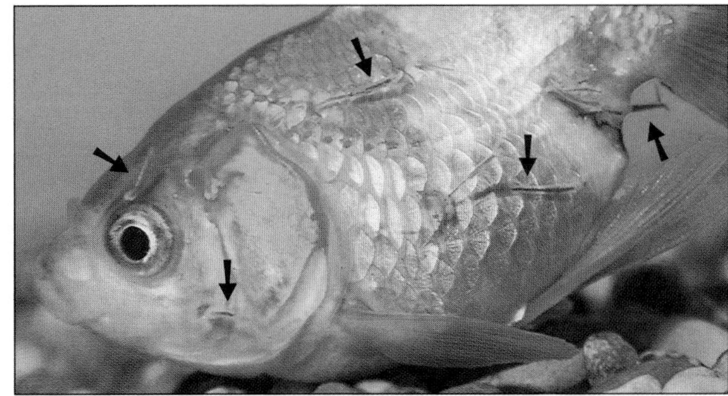

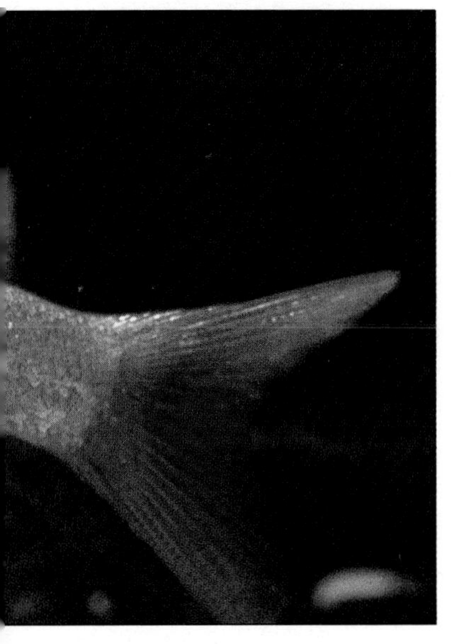

of parasites may need repeating after a week or two. During this process, keep the fish carefully but firmly restrained in a soft, damp cloth, and only keep it out of the water for a few minutes at a time.

If fish species or invertebrates susceptible to an organophosphorus-based remedy are present, be sure to remove them to a separate container while the pond or aquarium is being treated, and allow the active ingredients to dissipate for at least 10 days at summer temperatures – perhaps longer for some invertebrates. (See the manufacturer's instructions for use). While being maintained separately, inspect such fish for any signs of adult *Lernaea*, and remove these as described above.

The life cycle of anchor worm (*Lernaea*)

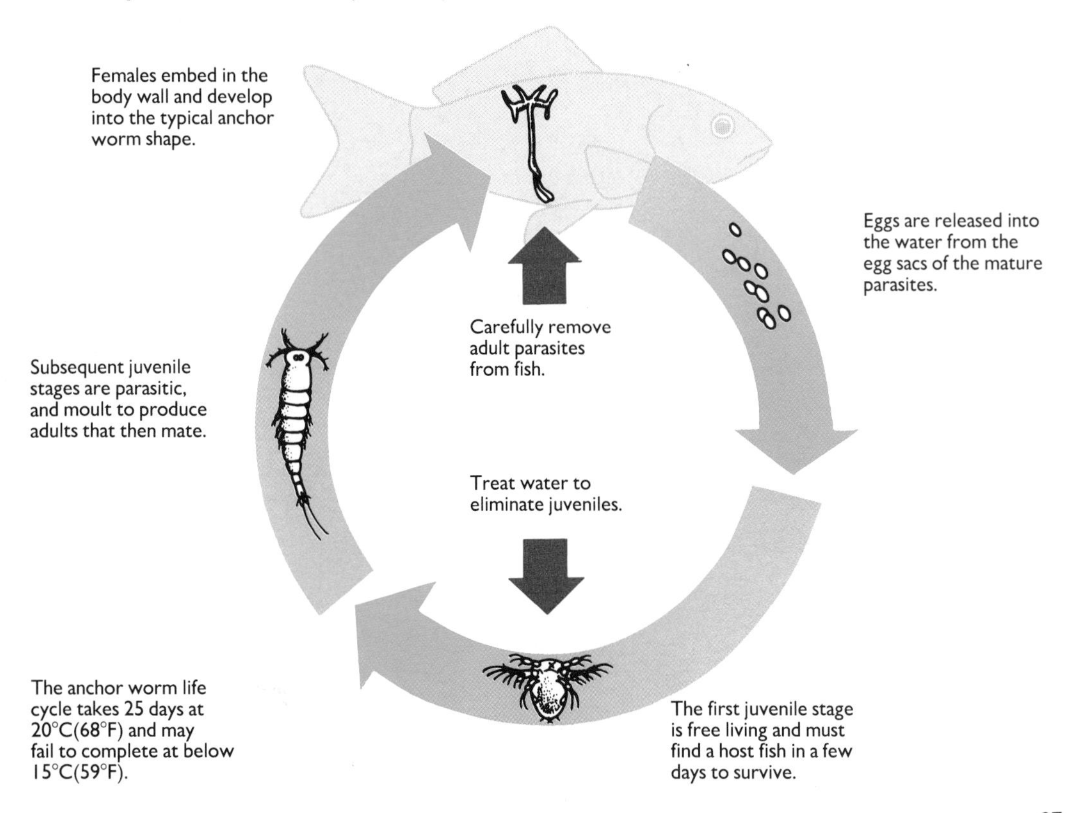

The anchor worm life cycle takes 25 days at 20°C(68°F) and may fail to complete at below 15°C(59°F).

BLOOD PARASITES

Caused by

Protozoans, such as *Trypanosoma* and *Trypanoplasma*, and the digenetic fluke, *Sanguinicola*.

Obvious symptoms

Noticeable symptoms are usually lacking, although heavy infestations can result in anaemia, listless behaviour, emaciation, 'pop-eye' and, in the case of *Sanguinicola*, extensive gill and kidney damage. Mortalities have been reported in ponds.

Occurrence of the disease

Since the blood protozoans require a leech vector (i.e. a carrier) to transmit them from fish to fish, and since *Sanguinicola* must undergo a period of development in a freshwater snail, these parasites rarely cause problems in aquariums. Even in pond situations, these parasites probably remain undetected unless moribund or dead fish are subjected to detailed post-mortem examination.

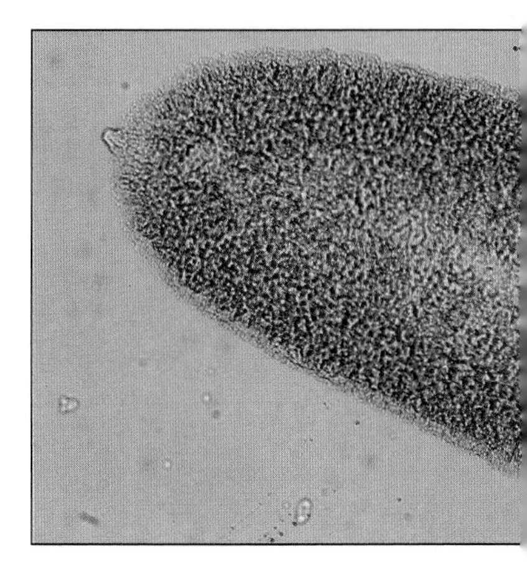

Above: A specimen of *Sanguinicola inermis*. These flukes are only 1mm(0.04in) long and are thus difficult to spot during post-mortem examinations.

The life cycle of *Sanguinicola*, a blood parasite.

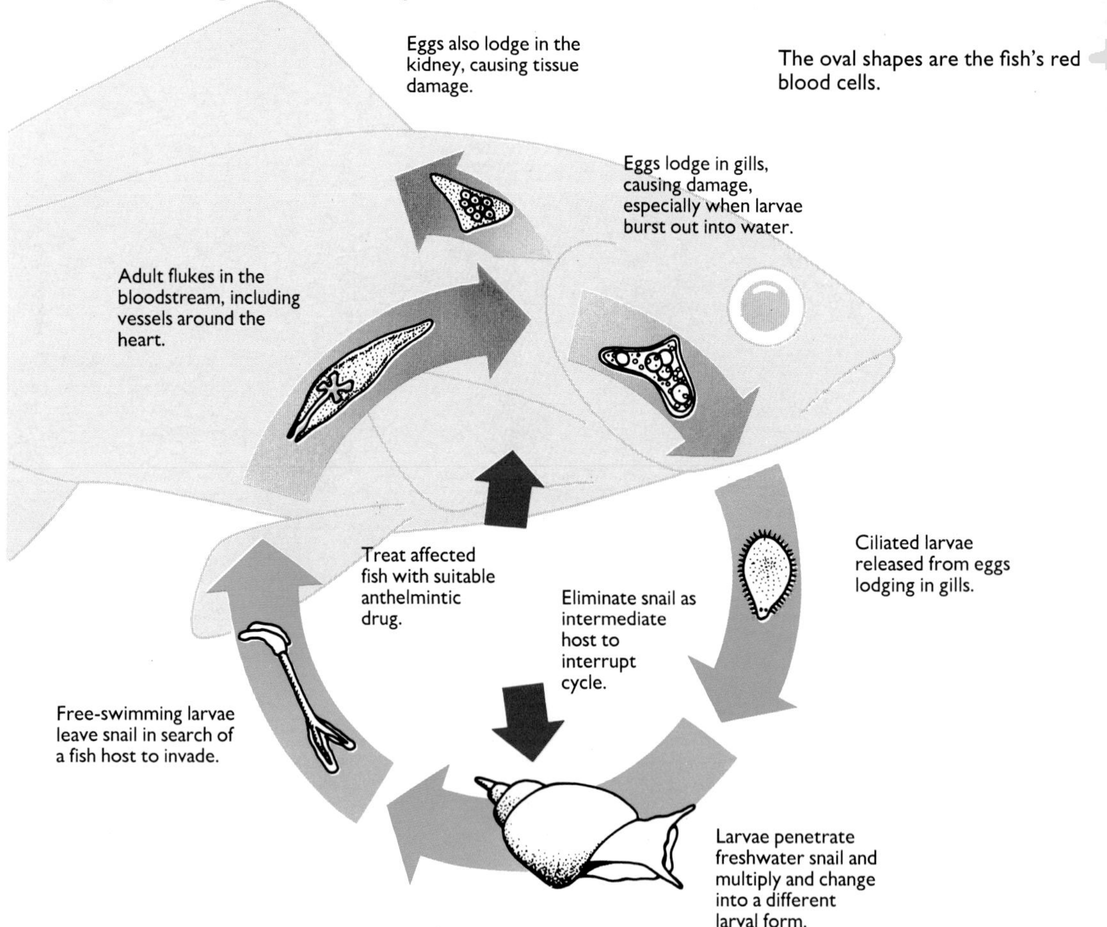

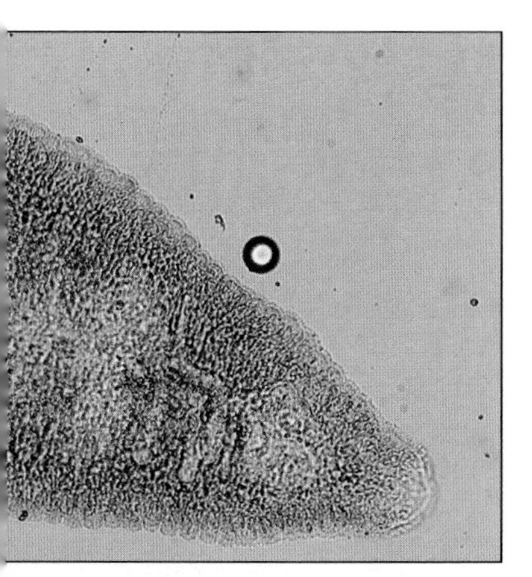

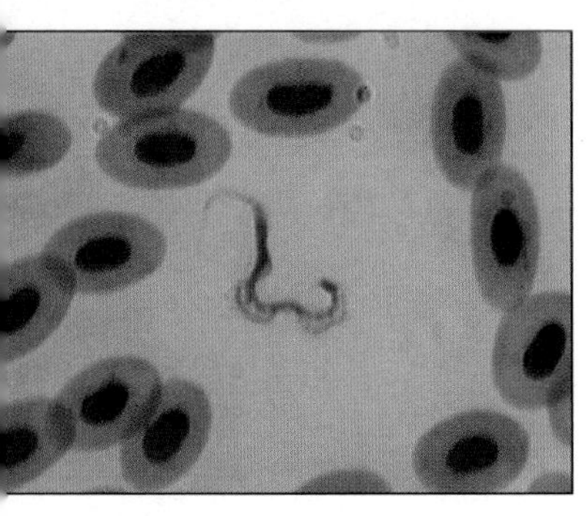

Treatment and control

There are few reliable treatments available to counteract infestations with blood protozoans, although some success has been experienced by using an injectable form of the anthelmintic drug praziquantel to treat fish infected with *Sanguinicola* (see Chapter 7). The best means of control is to eliminate the leech vector and snail intermediate host for these parasites from the pond and aquarium. The control of leeches is dealt with on page 124, and snails can be eliminated by draining and drying or by careful use of one of the available molluscicides.

Above: A sample of fish blood with a tiny eel-like protozoan *Trypanosoma* shown in the centre. Such blood parasites usually measure about 10-20 microns long.

Right: In the absence of other symptoms, listlessness and an emaciated appearance may point to infestation with blood parasites.

WHAT IS A PARASITE?

Parasitism as a way of life is just one of a huge range of naturally occurring animal (and plant) associations. In general terms, it is an association between different species in which one, the *host*, is indispensable to the other, the *parasite*, while the host can survive quite well without the parasite. In fact, the parasitic way of life involves a whole spectrum of associations, ranging from the very close to the very loose. Not all parasites are parasitic for their whole life, for example, and many spend different phases of their life cycle in very different hosts.

Parasites, of course, usually obtain their food from their host, which can cause problems for the host if it is fed a poorly balanced diet and/or if particularly large numbers of parasites are present. Normally, parasites do not kill their hosts, since this would be akin to suicide. However, under some circumstances, the delicate host-parasite balance can be upset, and result in the death of the host. Such situations can occur in an aquarium or pond from time to time.

Because the parasitic way of life is very hazardous, adult parasites usually produce very large numbers of eggs or larvae. This helps to ensure that at least one or two will be successful in finding a suitable host. Within the confines of a pond or aquarium, where host to host transmission is made easier by high stocking levels, this can result in a rapid build-up of parasite numbers.

Parasitism is a natural phenomenon, an integral part of the 'cycle of life'. However, when parasites cause problems for the fish hobbyist, it is usually the result of poor pond or aquarium management.

COTTON-WOOL DISEASE OR MOUTH 'FUNGUS'

Caused by
Usually the *Flavobacterium* bacterium (formerly known as *Flexibacter*)

Obvious symptoms
Initial signs of the onset of the disease are usually off-white marks around the mouth or on the fins or body of the fish. As the infection spreads, typical white cotton-wool-like tufts appear in the mouth region, with reddened ulcers on the body and frayed fins. In scaleless fish, a reddened edge may occur to any ulcers, and most affected fish – especially livebearers – often exhibit typical 'shimmying' behaviour, go off their food and appear very thin.

Below: Despite its appearance, mouth fungus is caused by the *Flavobacterium* bacterium, which is especially common in warm organically rich water. *Flavobacterium* also causes finrot (see page 106)

Occurrence of the disease
This disease is especially common in newly imported freshwater fish or freshwater fish in poor condition because of incorrect pond or aquarium care. A sudden change in water quality or generally unsuitable water conditions may trigger off the disease in previously healthy fish. Overstocking and infrequent water changes may also bring on an attack of cotton-wool disease.

Treatment and control
During the early stages, the disease is likely to be largely confined to the outside of the fish. At this point it is possible to treat the condition using an aquarium antibacterial or a phenoxyethanol-based remedy. In persistent cases or where established infections have invaded the internal organs, antibiotic or similar antibacterial treatment is necessary, as described in Chapter 7.

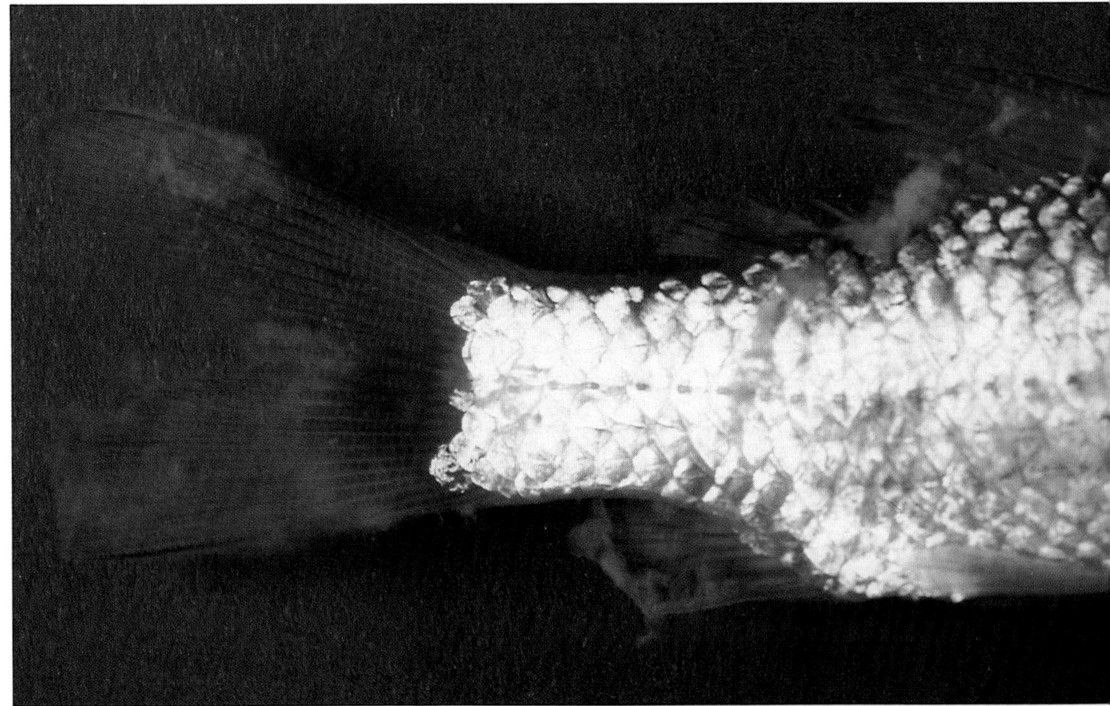

Right: A prepared slide showing long, thin *Flavobacterium* bacteria. These are up to 12 microns long and move with a smooth gliding and flexing motion.

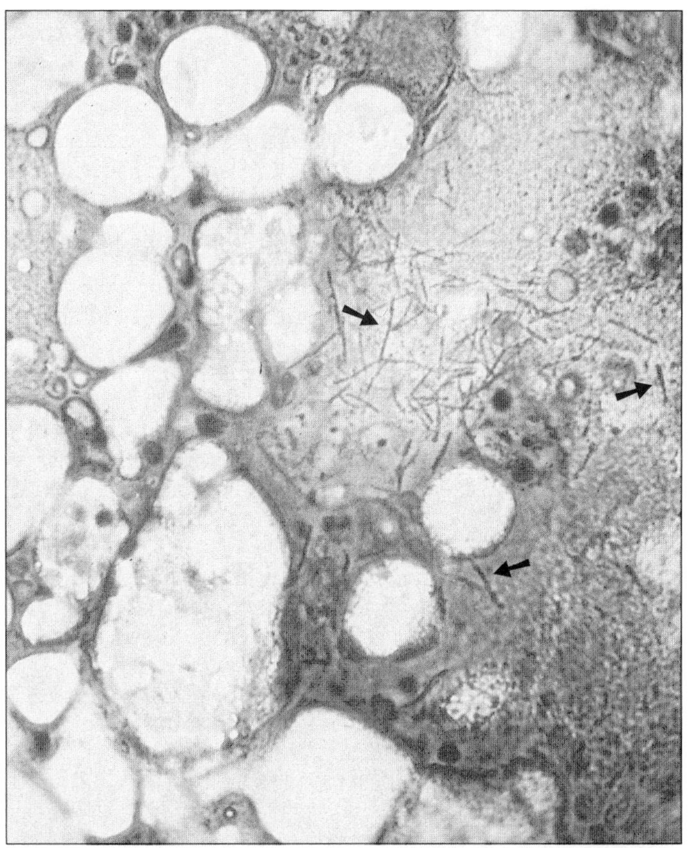

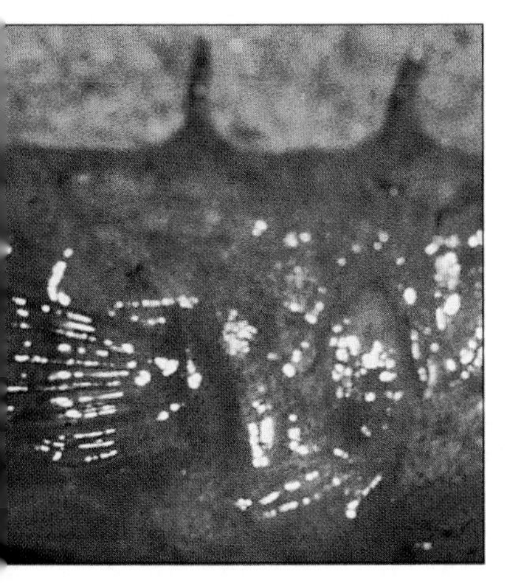

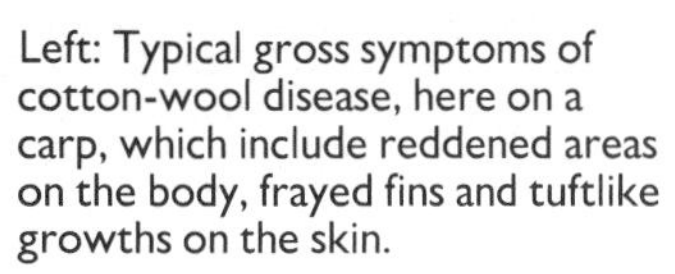

Left: Typical gross symptoms of cotton-wool disease, here on a carp, which include reddened areas on the body, frayed fins and tuftlike growths on the skin.

Right: Each tuftlike growth of cotton-wool disease contains many long, slender bacteria. Prompt treatment is vital before the internal organs are infected.

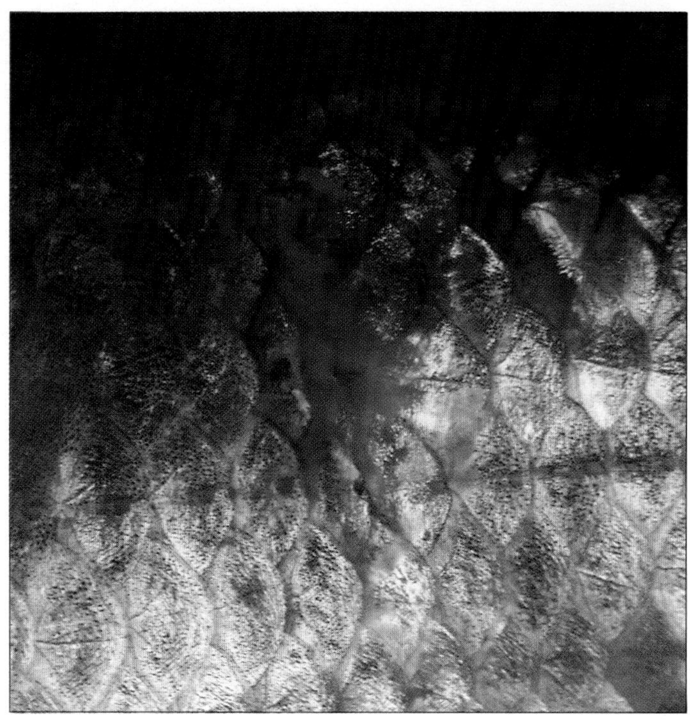

DEVELOPMENTAL PROBLEMS AND HEREDITARY DISEASES

Above: A male guppy with a skin tumour and small Siamese twin attached to the belly. The factors responsible for such events are not completely known.

Caused by

Crossbreeding of certain strains of fish can give rise to fish with an abnormal appearance. Various types of environmental factors, including pollution with metals and pesticides, unsuitable temperatures and low oxygen levels, can also influence the development of eggs and young fish. Some tumours may be inherited from parent fish, and certain deformities can be nutritional in origin – see page 130.

Obvious symptoms

Symptoms vary tremendously, from abnormalities in coloration and unusual fins to missing eyes, deformed jaws, shortened gill plates and loss of swimbladder (producing so-called 'belly slides'). Siamese twins is another manifestation of this sort of problem.

Occurrence of the disease

These symptoms may occur when fishkeepers are crossing certain strains of fish, or when the eggs or fry are kept in unfavourable water conditions. Fluctuating temperatures can cause spine deformities in fish fry, for example. Siamese twins is a common abnormality of livebearing fish, one of the twins usually being very much smaller than the other. Adding chemicals to the water containing eggs or fry may affect their development, as may the presence of heavy metal toxins, such as copper, or other pollutants. (See Chapter 3 for further information on water pollution.)

Below: Rainbow fish with kinked spine. This specimen also has tail rot.

Treatment and control

Grossly affected fish, or fish which appear to be suffering, are best painlessly destroyed (see page 200). Selecting alternative parent ('brood') stock and ensuring that the eggs and fry are kept in their preferred water conditions will help to control this type of problem. Treat eggs and fry only with the correct dose of reliable remedies and never expose them to toxins such as heavy metals. Condition all new tapwater for use with fish eggs and fry with a reliable water treatment. Avoid fluctuating temperatures during egg and fry development, and rear the eggs and young fish in the best possible conditions, particularly with regard to water chemistry and diet.

'Man-made' varieties

Some fish have been selectively bred in order to create unusual body shapes, examples being various round-bodied strains of goldfish (e.g. lionheads, orandas), and mollies ('balloon' mollies) and red parrot cichlids (deformed backbone and head). Others are injected with dyes, such as the 'painted' glassfish. The ethics of producing some of these unnatural strains and injected-colour variants has been questioned.

Below: Juvenile common goldfish with shortened gill covers. The fish appeared otherwise healthy and survived to old age.

DROPSY AND MALAWI BLOAT

Caused by
Bacterial and/or viral infections, also metabolic or nutritional disorders.

Obvious symptoms
Markedly swollen belly, protruding scales, reddening at vent or base of fins, ulcers on body, and long, pale faecal casts. Affected fish may go off their food and show darkening of coloration, pale gills and a 'pop-eyed' appearance. Fluid may accumulate in the body cavity, with discoloration of internal organs.

Occurrence of the disease
Dropsy may occur in fish which are in poor condition for some other reason or it may affect a small number of individuals in an otherwise healthy and well-maintained pond or aquarium.

Right: A build-up of fluid in the body can result in the swollen-eyed appearance often associated with dropsy, here in a koi. Such symptoms may subside with time.

Treatment and control

Because of the uncertain causes of this condition, precise treatment can be difficult. The best strategy is to remove affected individuals to a separate tank and give them the best possible food, ideal water quality conditions, etc. If their condition does not improve, treatment with a broad-spectrum antibiotic or similar drug may be warranted (see Chapter 7 for further information).

Left: This photograph shows the typical 'pine-cone' effect of a fish suffering from dropsy. This may not be infectious, but affected fish are best isolated.

Below: *Labeo bicolor* showing signs of dropsy. It is possible for a veterinarian to draw off excess fluid from within the body cavity using a fine syringe.

FINROT

Caused by

Usually various bacteria such as *Aeromonas, Pseudomonas* and *Flavobacterium* (formerly *Flexibacter*).

Obvious symptoms

Split, ragged or stumpy fins, often with a white edge to them. Concurrent infections with cotton-wool disease (caused by *Flavobacterium*) may arise (see page 100).

Occurrence of the disease

Finrot invariably occurs in fish that are in poor condition for some reason. Recent importation, rough handling, fighting – especially 'fin nipping' – overcrowding, incorrect water conditions and poor feeding will all predispose fish to finrot. Fancy coldwater fish with long trailing fins may develop finrot if the water temperature falls very low (below 10°C/50°F), or if they are left in a garden pond over the winter. (See Chapter 3 for more on temperature effects.)

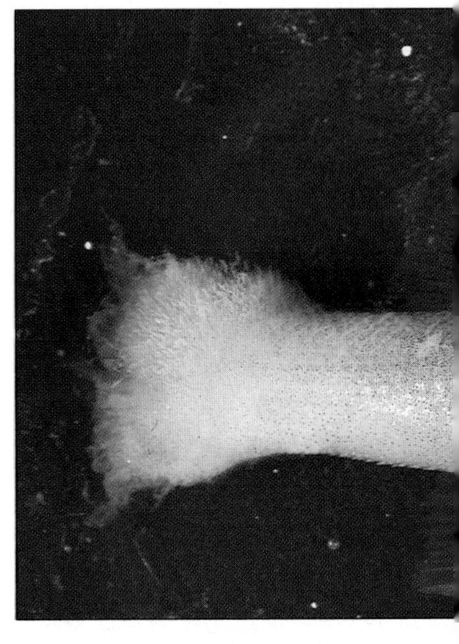

Above: Typical signs of severe finrot on a freshwater perch. If left untreated, such outbreaks can lead eventually to a systemic bacterial infection that will kill the affected fish and put others at risk.

Left: The bacteria responsible for finrot are common in aquarium and pond water, and like many potential pathogens, they form a constant threat to the fish.

Below: Fin damage caused by finrot, although fin nipping produces a similar result. With prompt treatment, damaged tissue will regrow in a few weeks.

Treatment and control

Using a suitable proprietary aquarium antibacterial usually cures most outbreaks. Adding some aquarium salt to the water may be useful in preventing finrot in some freshwater fish, such as livebearers that normally live in brackish water environments (e.g. guppies and mollies). Antibiotics or similar antibacterials may be used to control persistent cases; add the drug to the water of a treatment tank as described in Chapter 7.

It is important to emphasize that correct care usually prevents finrot developing in the first instance.

Right: Finrot often occurs with other diseases, as here in this fancy goldfish, and can indicate generally poor environmental conditions in the pond or tank.

Below: A female swordtail with its dorsal fin almost worn away by finrot, and clear signs that the infection is spreading to the body. Prompt treatment is vital.

FISH FUNGUS AND EGG FUNGUS

Above: Fungus on a goldfish. These and other coldwater fish are particularly susceptible to fungus in the early spring and following reproductive activity.

Below: The clear white threads of *Saprolegnia* fungus. Always treat these infections promptly or the disease will spread rapidly, perhaps with fatal results.

Caused by
Various species of aquatic fungi, including *Saprolegnia* and *Achlya*.

Obvious symptoms
Grey, brown or white cotton-wool-like growths or tufts on the skin and fins of freshwater and brackish-water fish. External fungal problems are rarely reported on marine fish. It normally begins as a small patch but can develop if left untreated, quickly killing the fish in some circumstances. Fungus may also attack fish eggs.

Occurrence of the disease
Fungus and fungal spores are quite common in aquatic environments and are particularly abundant where there is plenty of decaying organic matter. The infectious spores of fungus may transmit the disease from fish to fish. However, the layer of mucus that covers the skin of healthy, undamaged fish is normally a very effective barrier to the spores. If, for one reason or another, this mucus layer becomes damaged – following rough handling, fighting or spawning activity, for example – this will provide an opportunity for the fungus to gain a foothold. Fungus may also invade the lesions left by other diseases, such as white spot or ulcer diseases. A sudden change of temperature, unhygienic aquarium conditions and poor water quality may all predispose fish to fungus. In the case of fish eggs, fungus attacks the dead eggs and may spread to the adjacent healthy eggs, killing the whole batch.

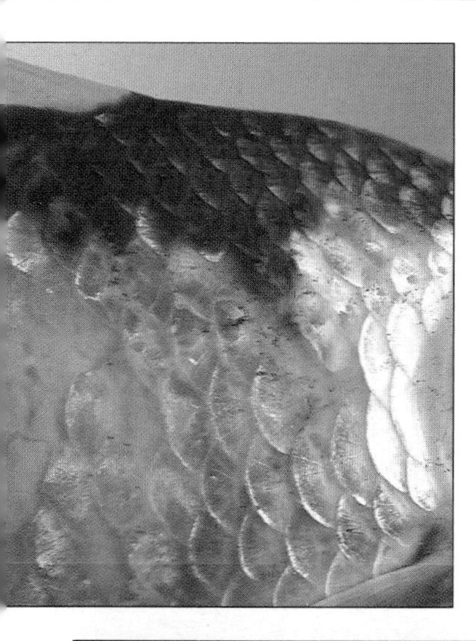

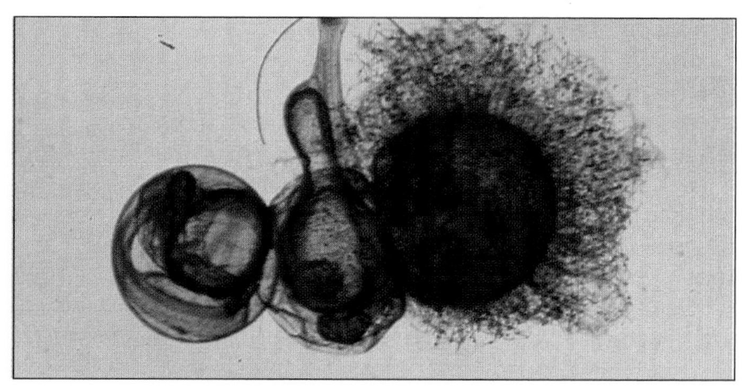

Above: Minnow eggs. The infertile egg on the right has been engulfed by fungus. The other two eggs are developing normally, with the central one starting to hatch.

Below: Developing fish fry next to badly fungused eggs. Unless prevented by prompt treatment, egg fungus will usually spread right across the whole batch.

Treatment and control

As soon as fish show any signs of fungus, treat them with a proprietary brand of fungus remedy. Treat lightly affected fish in the aquarium, but deal with heavily infected aquarium fish and all pond fish in an isolation tank.

To avoid egg fungus, remove all dead (opaque) fish eggs promptly and carefully from the batch using a fine pair of forceps or a pipette. Some proprietary brands of fungus remedy can be used to treat eggs, but be sure to check their instructions for use before proceeding. Some aquarists prefer to use methylene blue to prevent and treat fungus on fish eggs. This will not usually harm the fish, and is useful when some eggs, those of angelfish, for example, are incubated in the absence of the adult fishes.

To prevent fungus developing on fish and eggs, it is essential to identify and eliminate the underlying causal factors.

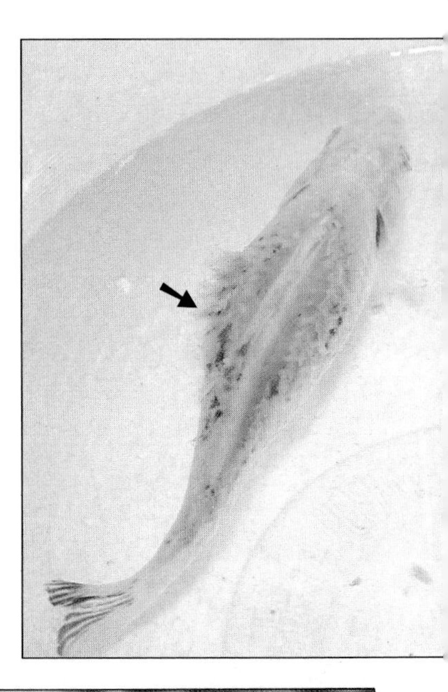

Right: The strands of fungus attached to this koi have taken on a greenish tinge from the accumulation of algal cells. The infection is clearly worsening.

Below: Fungal infections can progress swiftly, eventually encasing the entire fish, as here. Fungus also attacks fish that have died from other causes.

Right: Discus tending their brood of eggs, fanning fresh water and oxygen across them and removing any dead or fungused eggs to safeguard the healthy ones.

WHAT IS A FUNGUS?

There are approximately 50,000 species of fungus, and they are usually thought of as plants. However, they do not contain chlorophyll – the green photosynthetic pigment of plants – and cannot manufacture their own food. Some fungi are saprophytes, feeding on dead organic matter and thus helping bacteria to break down and release nutrients for recycling. Others are parasites of animals or plants, and some are an important source of antibiotics used in the treatment of bacterial disease of animals and man.

Fungi range in size from tiny single-celled yeasts, mildews and rusts (often seen on crops) to the familiar mushrooms and toadstools in damp fields. Fungi are widespread in damp terrestrial habitats and in aquatic habitats. The saprolegnia 'group' of water moulds will be familiar to most fishkeepers as the cause of fish and egg fungus. These fungi are made up of thin threads, or hyphae, which weave together to form an obvious fungal mat, or mycelium. These threads often need to gain access through damaged skin, but once inside can then penetrate deep into tissue using special enzymes.

Fungi can reproduce by asexual or sexual means. Either process usually results in the liberation of spores, certain types of which can be quite resistant to adverse conditions.

It is important to note that the fish disease 'mouth fungus', although it looks like a typical, fluffy fungal infection, is, in fact, the result of an infection with the *Flexibacter* bacterium (page 100).

Below: These fungal strands, or hyphae, are about 20 microns in thickness. They intertwine to make the familiar fungal mat.

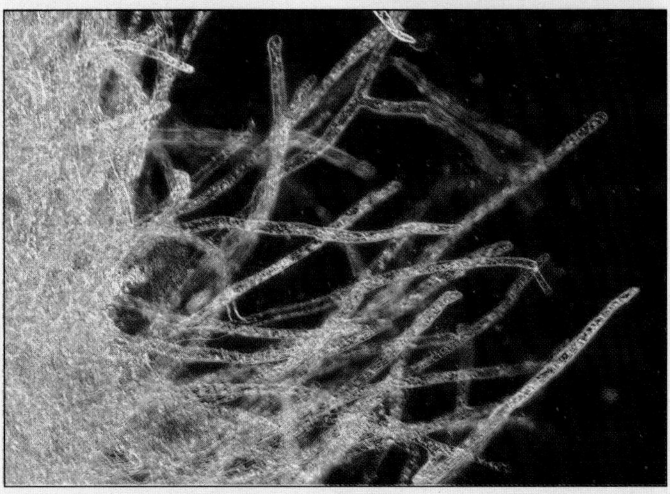

Below: The clublike ends to the fungal strands visible here contain many spores that may eventually infect other fish.

FISH LICE AND GILL MAGGOTS

Caused by
Various crustaceans (e.g. fish louse, *Argulus*; gill maggot *Ergasilus*).

Obvious symptoms
The fish louse (*Argulus*) is a flattened, disc-shaped crustacean measuring up to 10mm(0.4in) in diameter. It attaches itself to the skin and fins by means of twin suckers and feeds on blood by inserting its sharp mouthparts into the body of the fish. The intense irritation that results may cause heavily infested fish to scratch against rocks and even jump out of the water. Reddened lesions develop at the point at which the parasite is feeding, and these may become secondarily infected with fungus or bacteria.

Gill maggots (*Ergasilus*) are usually found attached to the gills, gill covers and inside the mouth. These parasitic crustaceans are usually several millimetres long and the common name refers to the adult females with their prominent, whitish, 'maggot-like' egg sacs. (The male does not become a parasite, as the life cycle diagram shows). Heavy infestations with *Ergasilus* can cause severe gill damage, emaciation, anaemia and even death.

Occurrence of the disease
The life histories of these parasites are markedly affected by temperature, and they are most often a problem in garden ponds in the summer months. They do not often cause problems in home aquariums, although they may be found on newly imported fish.

Argulus may overwinter in a pond as eggs, juveniles or adults. When the eggs hatch, the juvenile parasites must find a host within a few days. Adult *Argulus* can live away from the fish host for as long as 15 days. Because of their bloodsucking feeding habits, they may transmit certain microbial infections between fish. Heavy infestations are especially dangerous to small fish.

Above: An isolated fish louse (*Argulus* sp.) viewed from above. Mature specimens reach about 10mm(0.4in) across. Note the black eyes and feathery legs.

Right: Low-level infestations of fish lice probably cause little harm, but they can rapidly build up to large numbers in warm pond water during the summer months.

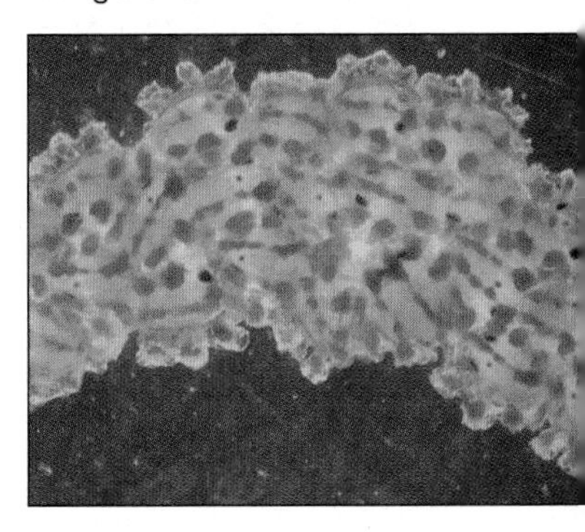

Above: Section of egg-mass, produced by the fish louse. Fish lice lay their eggs in thin strips a few centimetres long on submerged hard surfaces, including the glass of an aquarium.

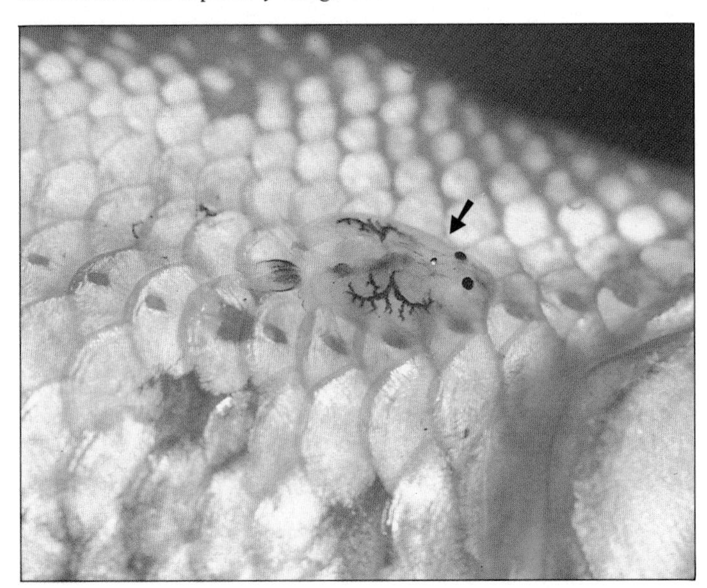

Left: Although most obvious when attached to the fish, here shown on a goldfish, *Argulus* can spend quite long periods – perhaps two weeks or more – free swimming in the pond or aquarium water.

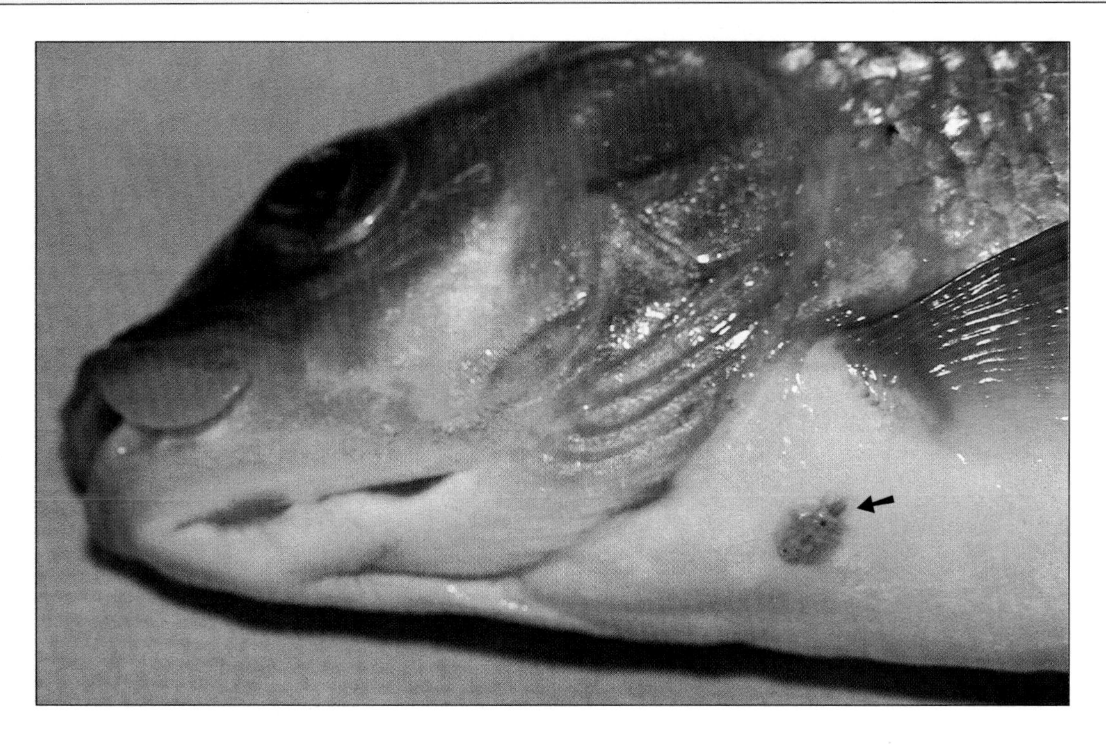

The life cycle of the fish louse (*Argulus*)

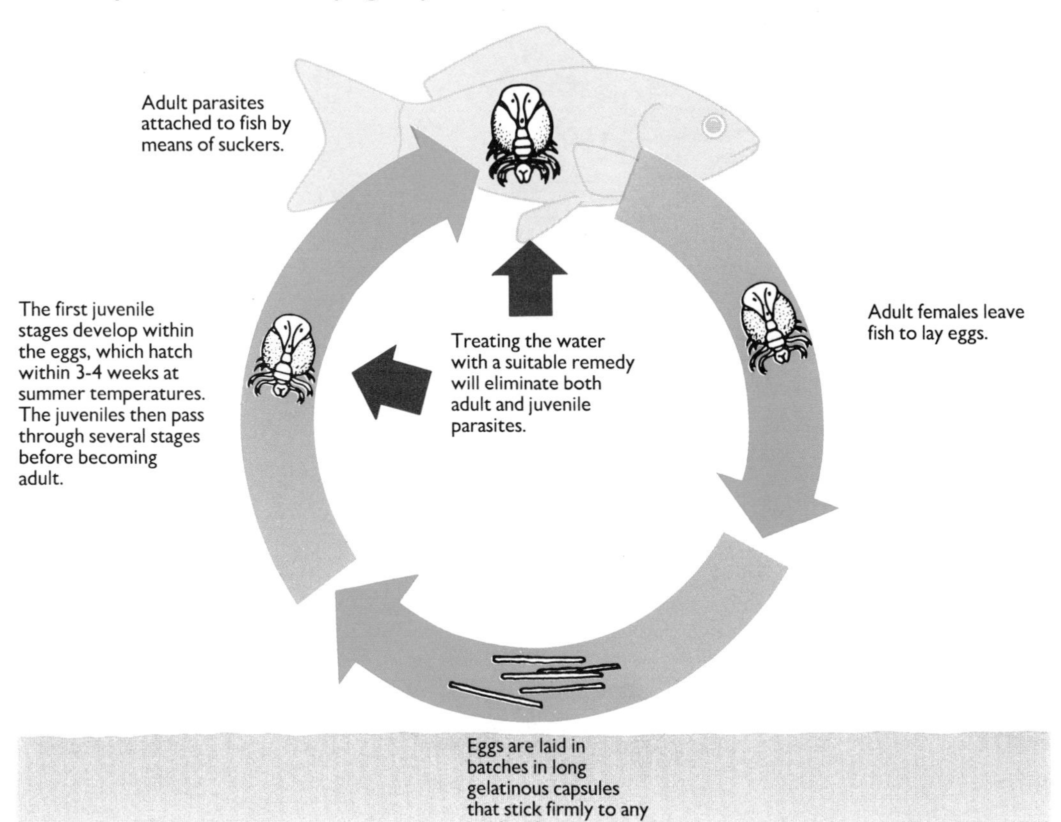

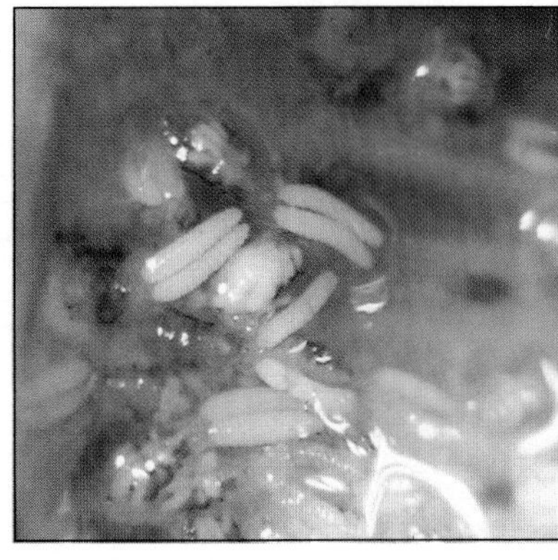

Above: These white egg sacs are characteristic of mature female copepod crustaceans and have earned *Ergasilus* the appropriate common name of 'gill maggot'.

Left: The gills of a bream infested with *Ergasilus*. When present in large numbers, these crustacean parasites can cause severe gill problems for fish, even death.

The life cycle of the gill maggot (*Ergasilus*)

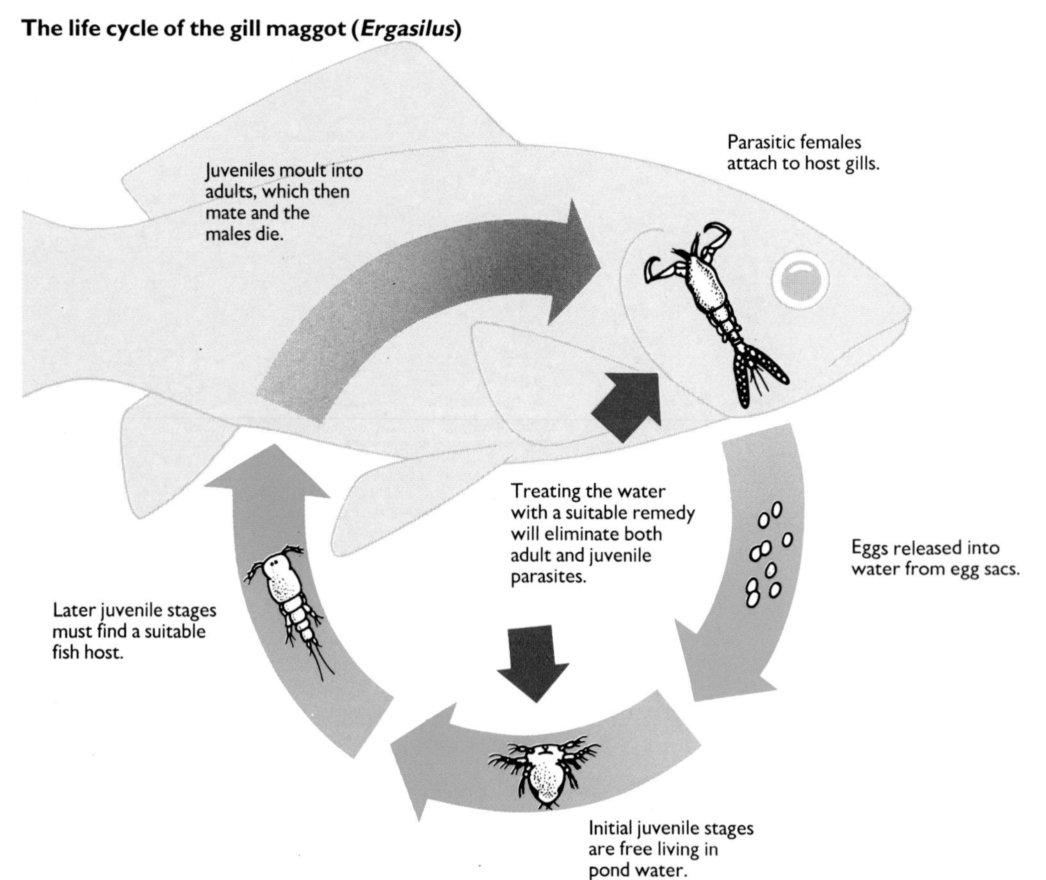

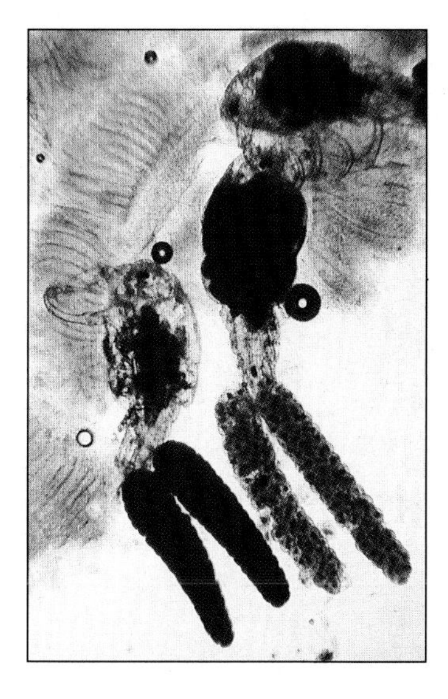

Above: Viewing *Ergasilus* through a low-power microscope shows that the antennae are modified to form gripping 'jaws' for attachment to the gill filaments of the host.

In *Ergasilus*, the male has a relatively short lifespan and it is the female (with egg sacs) that is most often seen attached to the fish. When the eggs are laid, they hatch after a few days and give rise to juvenile stages that seek out and attach themselves to fish. After mating, the males die, leaving the mature females attached to the fish, where they can live for some time – perhaps a year.

Treatment and control

The adult stages of *Argulus* and *Ergasilus* (and the juvenile stages of most crustacean parasites) are susceptible to treatment with organophosphorus insecticides, such as metriphonate (see Chapter 7). Apply the treatment to the infected pond or aquarium; one, or possibly two courses of treatment should suffice. Some fish, such as orfe, rudd and piranhas, and marine invertebrates are very sensitive to this type of chemical. If creatures sensitive to these remedies are present, be sure to remove them while the pond is being treated and allow ample time for the chemical to dissipate before returning them. This normally takes a week or two in warm alkaline water, but much longer under cool acid conditions. While any sensitive fish are in a separate holding container, you can treat them with a 30-minute bath of potassium permanganate before reintroducing them into the pond or aquarium. (See Chapter 7 for guidance on the strength of solution required.)

WHAT IS A CRUSTACEAN?

Crustaceans are members of the phylum Arthropoda, a vast assemblage of jointed-limbed invertebrate animals which share the possession of a tough external skin, or 'exoskeleton'. About a million species of arthropods are known, which represents approximately 75 percent of all the animal species so far described to science.

Over 35,000 species of crustaceans (class Crustacea) are known, and many of these are free-living aquatic species that 'breathe' by means of gills. Male crustaceans typically transfer sperm to the female, who then carries the eggs until they hatch. The larvae are often free-swimming members of the plankton, which must develop through several juvenile stages until they become adults.

Crabs, lobsters, crayfish, shrimps and 'water fleas' (e.g. *Daphnia* and *Cyclops*) are all crustaceans. Two groups of crustaceans – copepods and branchiurans – are also of particular importance as causative agents of fish diseases.

- **Copepods** (Copepoda) embody approximately 4500 species of typically free-living members of the plankton, in both fresh and salt water. The freshwater 'flea', *Cyclops*, is within this group. However, a few species have become parasitic on a range of aquatic animals, especially fish. Such parasites (e.g. *Ergasilus* and *Lernaea*) can have quite complicated life histories, with the juvenile parasites living in the plankton. The adult stages of some parasitic copepods (in particular the anchor worm, *Lernaea*) are so highly adapted to their parasitic way of life that they are scarcely recognizable as copepods – or even as crustaceans.
- **Branchiurans** (Branchiura) include about 75 species of parasitic crustaceans. The fish louse (*Argulus*) exhibits the typical branchiuran features of a flattened body with specialized biting and sucking mouthparts as parasitic adaptations.

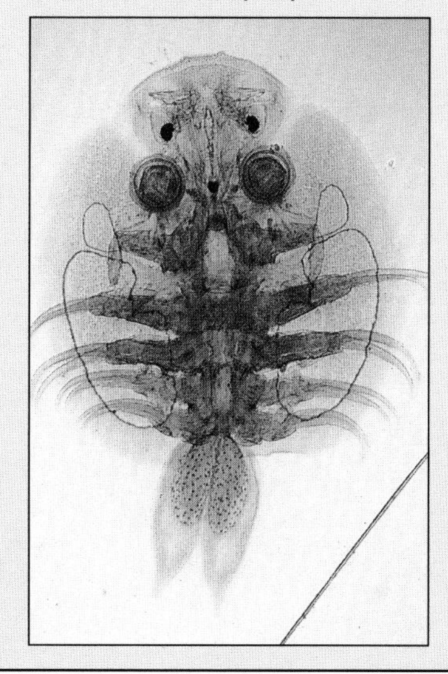

Right: In this close view of *Argulus*, note the two obvious suckers on the underside that are used for host attachment.

In addition, there are a number of less frequently encountered fish parasitic crustaceans, including the isopod *Livoneca*, a marine fish parasite that is a relative of the familiar woodlouse, and the amphipod *Laphystius*, another marine fish parasite related to the river shrimp.

FISH POX

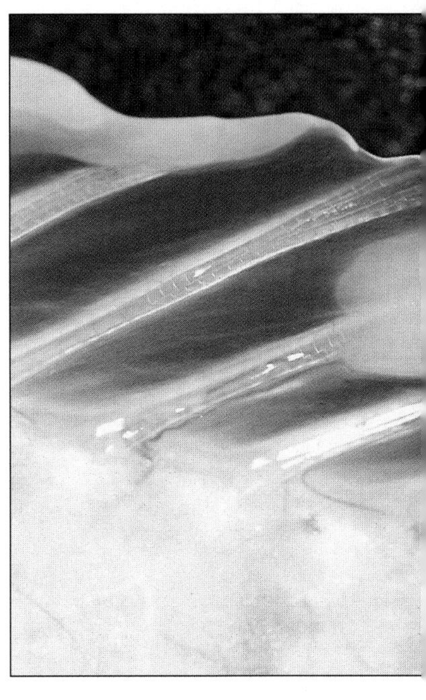

Caused by
A viral infection.

Obvious symptoms
White, grey or pink growths appear on the skin and fins. In many instances, an affected fish looks as if molten candle wax has been poured over the body. Occasionally, in extreme cases, the growths may become very pronounced and bear similar pigmentation to the surrounding skin. (See also *Tumours*, page 152.)

Occurrence of the disease
The disease most often affects coldwater aquarium and pond fish, especially koi. Fish pox often appears, develops to a certain extent and then subsides and disappears, sometimes to reappear at a later date. It is not very infectious and other fish in the same pond or tank may remain unaffected. It rarely causes any fish losses.

Treatment and control
There is no reliable treatment, although raising the water temperature by 5-10°C(9-18°F) sometimes eliminates the problem on a temporary basis. Since even badly affected fish do not appear to suffer, and since the infection is not markedly infectious, this disease does not really give any cause for concern. It is unsightly rather than dangerous. Nonetheless, avoid buying obviously infected fish from dealers.

Below: These pale waxy growths are typical of fish pox in goldfish or koi. It is sometimes confused with lymphocystis, which has a rougher, rasberry-like growth.

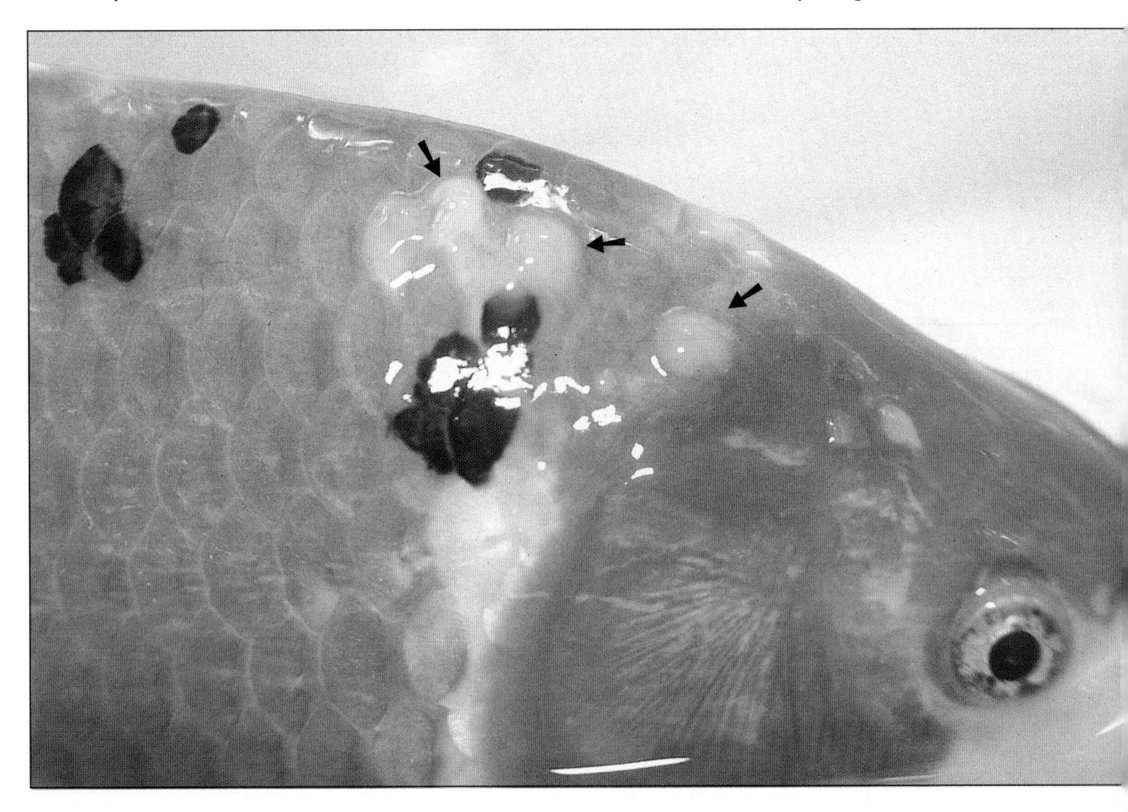

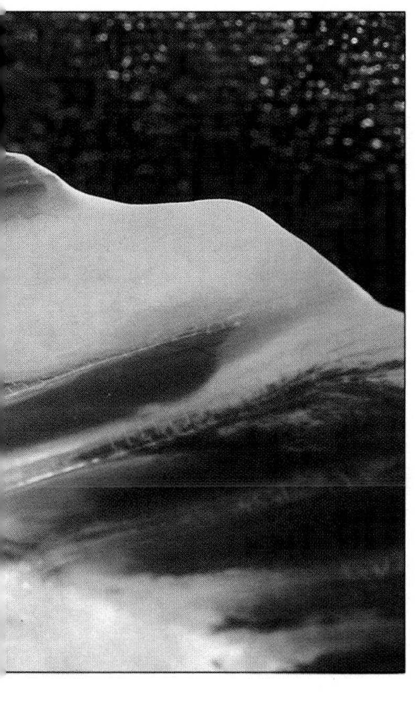

Above: The smooth growths of fish pox are unsightly rather than dangerous, which is fortunate because there is no reliable treatment for this viral disease.

Above: An extreme form of fish pox, with the growth taking on the coloration of the surrounding tissue. Even so, the fish was able to feed and behave normally.

WHAT IS A VIRUS?

A virus is a very small, very simple organism. Its structure is so simple that it is often considered to be on the borderline between a living organism and inanimate particle.

Viruses are so small – even smaller than bacteria – that they cannot be seen with an ordinary light microscope; a powerful electron microscope is necessary. Viruses often range in size from 10-500 nanometers (millionths of a millimetre).

All viruses are parasitic, and can live and reproduce only inside other living cells. As a result, they cause a range of diseases. In man, these include measles, influenza, rabies and poliomyelitis. Even the common cold is a viral infection. Viral infections that occur in fish include those that cause fish pox, lymphocystis and some tumours.

Because viruses live inside the cells of animals and plants, chemical treatment is generally out of the question, since killing the virus would also mean killing the host cell. Although certain drugs can alleviate some of the symptoms, the only effective way of controlling a viral infection is to isolate the affected individual from further contact with the virus and allow its own immune system to counteract the infection. Some viral diseases in man, such as smallpox and poliomyelitis, can be prevented by vaccination (i.e. stimulating an immune response to protect the individual against a specific virus). Unfortunately, there are viruses which can live for quite some time outside the body of the host, and some can survive freezing and even drying.

Typical virus shapes

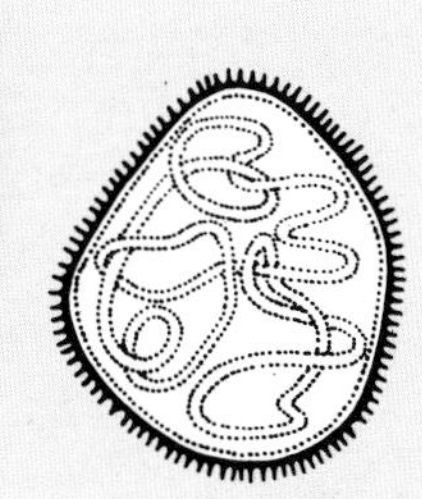

These diagrammatic representations illustrate the shape and form of various virus particles. Not to scale.

GILL DISEASE

Caused by

Gill problems may be caused by poor water conditions (e.g. ammonia or chlorine in the water) and/or by infections with certain fungi (such as *Branchiomyces*), bacteria, protozoans and monogenetic flukes (such as *Dactylogyrus*).

Obvious symptoms

Telltale signs include rapid gill movements, swollen gills and discoloured gill filaments with excess mucus. Fish go off their feed and lie motionless in the tank or gasp at the water surface.

Occurrence of the disease

This condition often occurs in newly imported fish kept in poor conditions, or in established tanks and ponds which are poorly maintained. Poor filtration and/or aeration, overcrowding and infrequent partial water changes may predispose fish to gill problems. On the other hand, carrying out total (rather than partial) water changes and using unconditioned tapwater may irritate the delicate gill membranes, making them more susceptible to infection.

Treatment and control

First investigate whether a water problem (e.g. a toxic level of ammonia) is the cause. In which case making a prompt 25-50 percent partial water change (using properly conditioned water) is a good initial treatment. If necessary, add an antibacterial treatment to a set-up aquarium; treat pond and marine fish separately in an isolation tank.

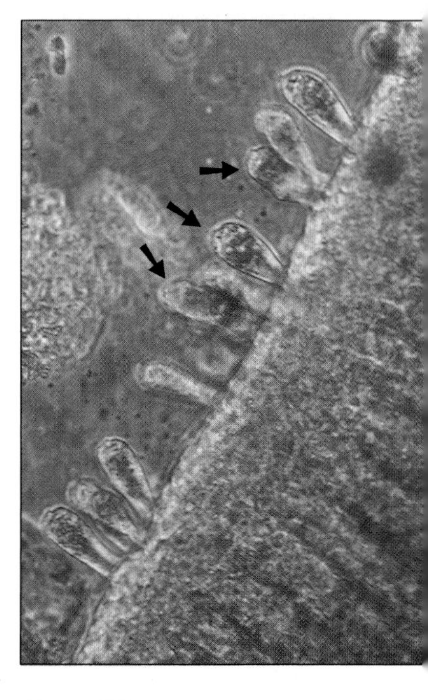

Above: Ciliate protozoans, *Apiosoma* sp., on a gill filament. Such parasites can block the gill surfaces if present in large numbers. Up to 50 microns long.

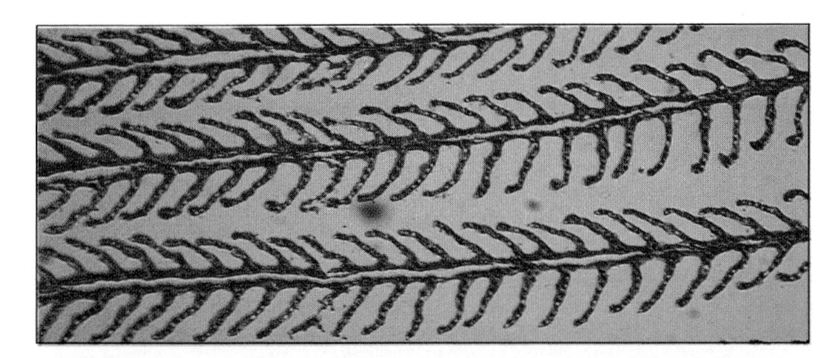

Left: A thin section through a healthy gill filament showing the waferlike lamellae that give each filament an enormous surface area for efficient gaseous exchange.

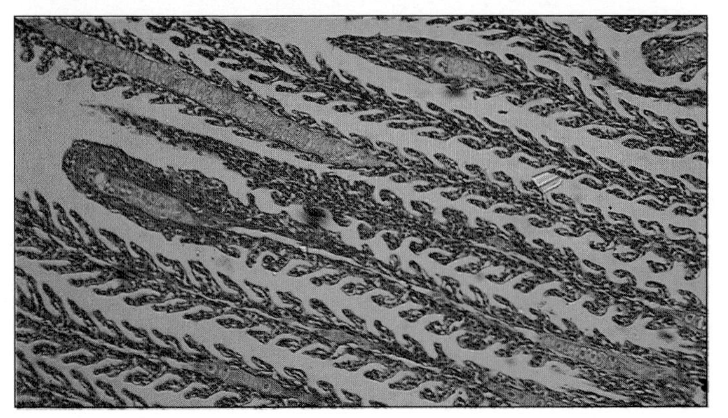

Below left: Infections and/or poor water quality can make the gill filaments and lamellae clubbed and thickened, seriously reducing their surface area, with consequent problems for the fish.

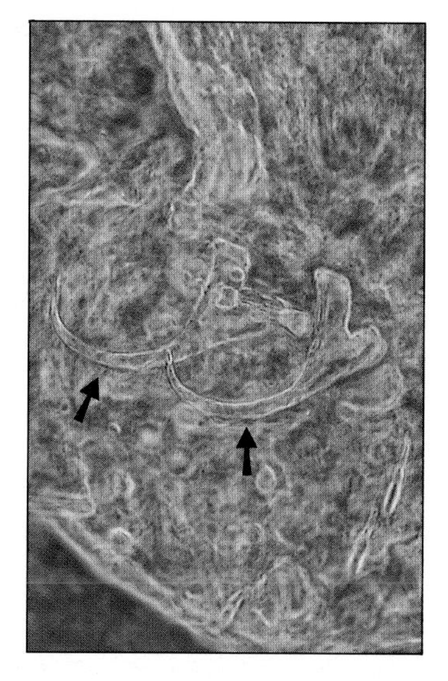

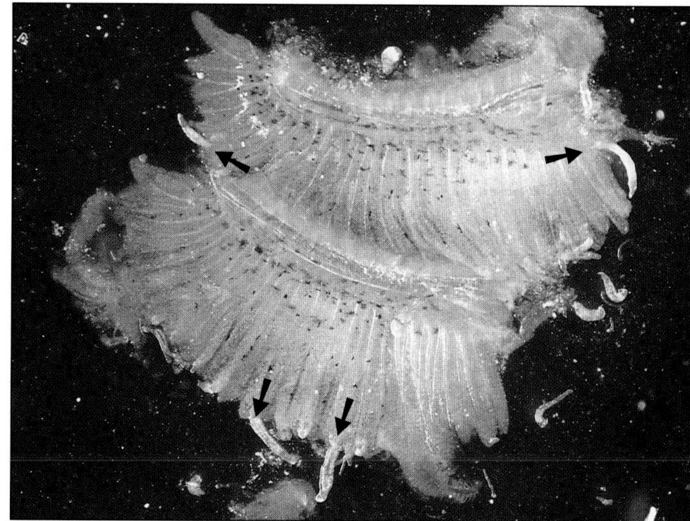

Above: This magnified view shows the hooked attachment organ that *Dactylogyrus* uses to grasp the gill filaments of its host.

Above right: A fragment of dissected gill clearly showing the *Dactylogyrus* flukes clinging to the filaments; each fluke is about 1-2mm(0.04-0.08in) long.

If an outbreak does not respond promptly to any of the above measures, or if it appears to be the result of gill parasites, use an organophosphorus insecticide such as metriphonate, formalin or a full course of copper treatment. These treatments can be toxic to invertebrates and/or some fish, however. Gill fungus, caused by *Branchiomyces*, is very difficult to treat. Carefully monitoring the water conditions and carrying out regular partial water changes will help to prevent this type of problem arising.

The life cycle of the gill fluke (*Dactylogyrus*)

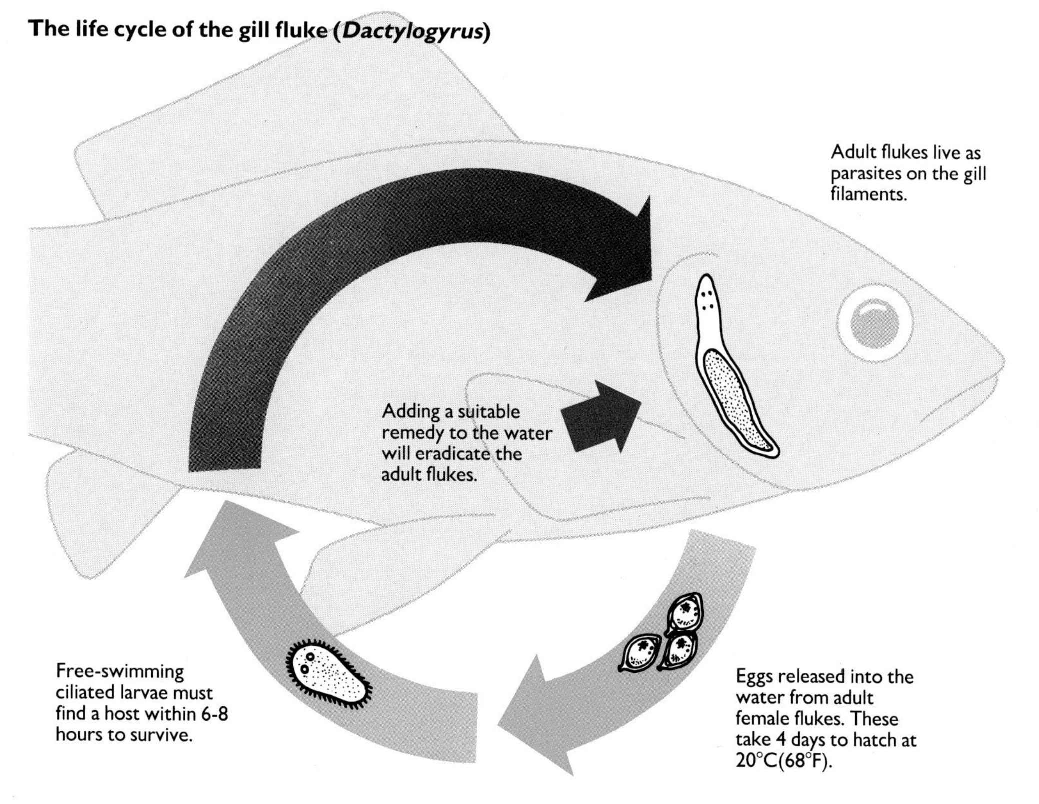

GLOCHIDIAL INFESTATION

Caused by
The larval stages of freshwater mussels, such as *Unio* and *Anodonta*.

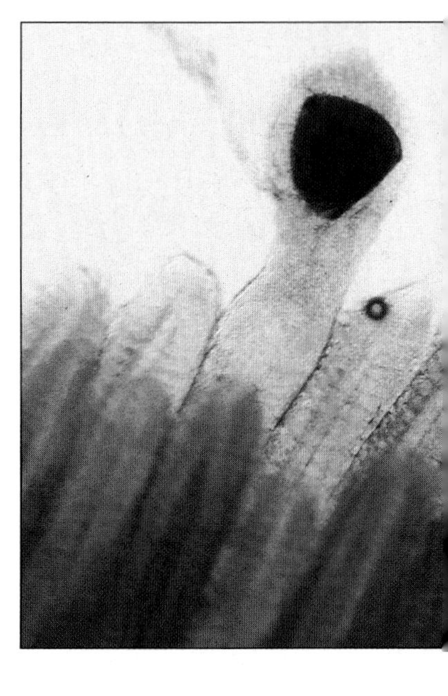

Obvious symptoms
Grey edges to the fins and/or gills are usually the most obvious symptoms, which can be quite pronounced in heavily infested fish. Detailed examination reveals the presence of tiny mussel larvae, or glochidia, which are less than 1mm (0.04in) in size.

Occurrence of the disease
This disease may occur in freshwater fish that have been kept with freshwater mussels. It is therefore generally restricted to pond fish (notably during the spring and summer), although coldwater aquarium fish are also affected. The tiny glochidia are expelled from adult mussels during the spring and summer, and must find a fish host before they can develop further. The larvae embed themselves into the skin, fins and gills of the fish and remain parasitic for several months. Eventually, fully formed miniature mussels fall away and begin a free-living existence. While attached to the fish, this parasite does not appear to do a great deal of harm, unless present in very high numbers.

Treatment and control
There is no reliable treatment for the glochidia while they are attached to the fish, but, fortunately, this condition does not usually warrant treatment. The surest way of preventing an infestation is to remove adult mussels from the aquarium and pond while they are shedding their glochidia during the summer period.

Below: An enlarged view of two glochidia, which appear as tiny replicas of the adult mussel and complete their development as parasites on fish during summer.

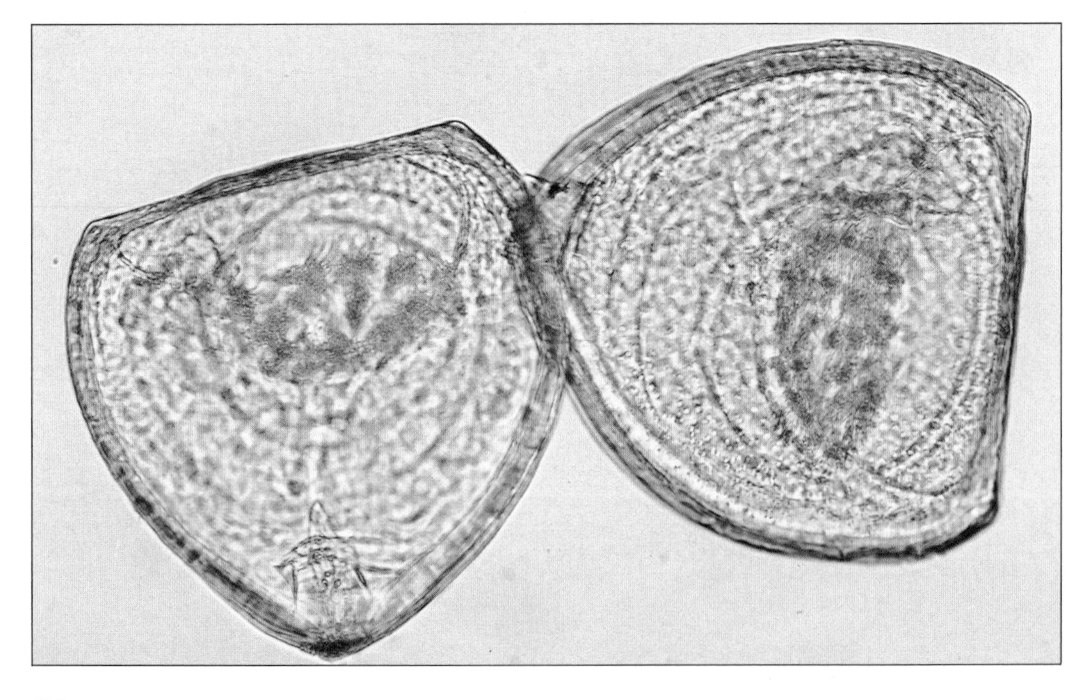

Below: A glochidium embedded in fish host gill tissue. Heavy infestations may cause symptoms of gill disease (see page 118).

WHAT IS A GLOCHIDIUM?

A glochidium (plural glochidia) is the larval stage of a freshwater bivalve mollusc. The molluscs are a large group of more than 100,000 species, examples of which are found in terrestrial habitats, as well as in fresh and salt water. Slugs, snails, limpets, octopus and squid are all molluscs, although it is the Lamellibranchia group of molluscs – the bivalves – that include the familiar freshwater mussels.

Bivalves (and there are over 8000 freshwater and marine species) are aquatic, sedentary or slow-moving molluscs with a flattened shell divided into two halves and joined by a hinge. They have large gills within their shell, and draw water into their body through an inhalant siphon 'powered' by tiny cilia. Bivalves feed by straining small particles from this flow, before returning the water to the outside world via the exhalant siphon.

Bivalves can be hermaphrodite or have separate sexes. Sperm from one mussel is usually drawn into the body of another mature individual through the inhalant siphon, thereby fertilizing the eggs within. The fertilized eggs may then take several months to develop, before tiny larval bivalves, the glochidia, are released in large numbers. Glochidia from the freshwater mussels *Unio* and *Anodonta* must then find a fish host on which they become external parasites while they complete their development.

As an example that redresses the balance, the bitterling (*Rhodeus*) lays its eggs inside freshwater mussels, where the eggs and fry develop, perfectly safe from predators.

The life cycle of glochidia from freshwater mussels

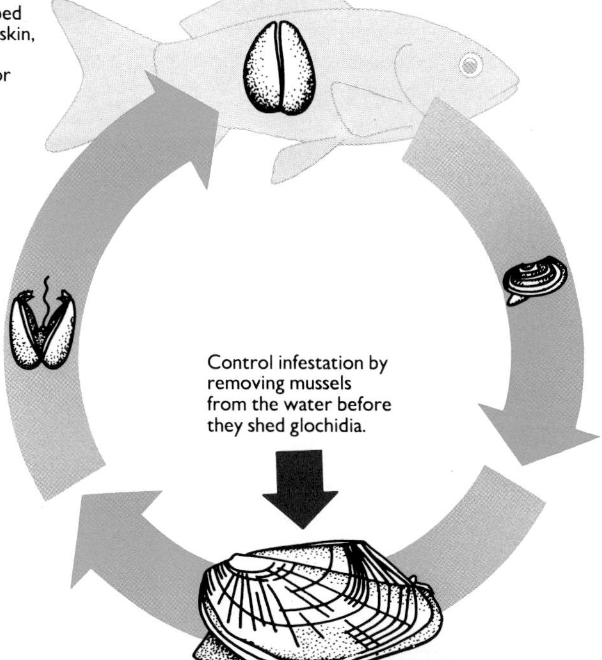

HOLE-IN-THE-HEAD DISEASE

Caused by

The underlying cause or causes remain unclear. Infection with flagellate protozoans (*Hexamita* or *Spironucleus*) have been implicated in some cases. Inadequate nutrition, bacterial infections, or poor water conditions may also be contributory factors.

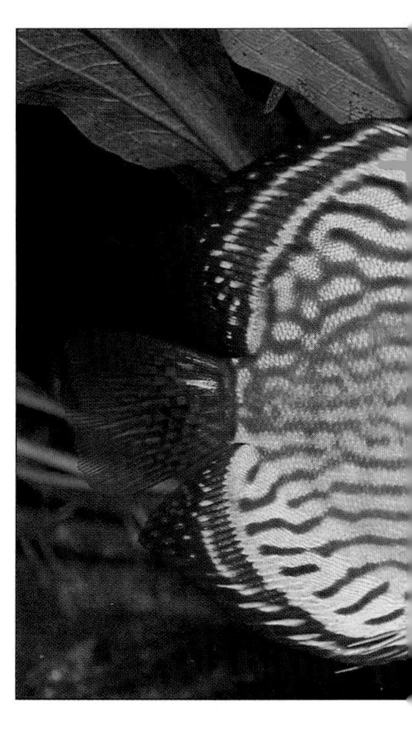

Obvious symptoms

Small holes appear in the body, especially the head region, which gradually develop into tubular eruptions. Very often, yellow, cheesy strings of mucus will trail from the lesions, leading some aquarists to believe that their fish are suffering from a 'worm infestation'. Affected fish often go off their food and develop a hollow-bellied appearance, with pale, stringy faeces. Lesions may also develop at the base of the fins and near the lateral line. Somewhat similar symptoms occur in a disease known as 'Head and Lateral Line Erosion' (HLLE) that affects marine fish, notably surgeonfishes (Acanthuridae) and marine angelfishes (Pomacanthidae).

Occurrence of the disease

Hexamita often exists as a low-level infection in the intestines of a variety of coldwater and tropical fish, notably affecting cichlids – such as discus, angelfish and oscars – and gouramis. Such infestations probably do the fish little harm. However, various factors, such as overcrowding, low oxygen levels, unhygenic conditions, changes in temperature and poor diet, may cause the parasites to multiply and then develop the acute symptoms described above. HLLE may be a complex disease syndrome related to poor diet (lack of vitamin C), as well as environmental factors.

Right: An advanced case of hole-in-the-head disease. This discus had been preserved in formalin and submitted for laboratory examination (see pages 78–80).

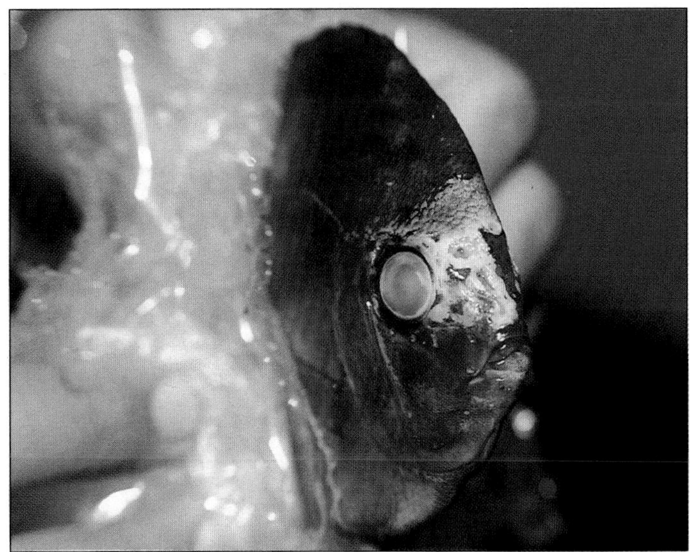

Above: The early stages of hole-in-the-head disease begin with small pale lesions in the head and sides of the fish, as is visible on this discus.

Left: A range of fish species may suffer from hole-in-the-head disease, although cichlids (as shown here) and gouramis are most often affected by this infection.

Treatment and control

The best way of treating hole-in-the-head disease is with medicated food. Unfortunately, affected fish often stop feeding and it is difficult to mix medicated food on a small scale. Fortunately, there are a few drugs which may be added to the water of the tank. Among these are dimetridazole and metronidazole, both of which are available only with a veterinary prescription in certain countries. A local veterinarian should be able to assist you in the use and application of these drugs, and it is likely that several repeat treatments will be needed for a successful outcome. Further details of these treatments are provided in Chapter 7. Ask at your local aquarium shop about the availability of proprietary treatments to combat hole-in-the-head disease.

For long-term control, it is important to quarantine all new stock, give appropriate preventative treatment and eliminate the factors which bring on the disease.

In cases of the marine form of this disease (HLLE), successes have been reported using vitamin C supplementation to the diet and the elimination of any environmental stressors. Stray voltage in the water (caused by electrical equipment that has not been properly grounded) has also been suggested as a possible cause.

LATENT INFECTIONS

It is seldom realized that most apparently healthy fish are usually carrying a range of potential fish pathogens in or on their bodies. These low-level, or 'latent', infections probably do the fish very little harm, but their importance to fish hobbyists cannot be overemphasized. It means that even within the healthiest and best-kept pond or aquarium there lurk pathogens waiting to cause problems for the less-than-vigilant hobbyist.

For the most part, caring for the fishes and their environment correctly will allow the natural immune reactions of the fish to keep these pathogens in check. If environmental conditions deteriorate, however, this may reduce the resistance of the fish and result in an outbreak of disease. Unfortunately, even after successfully treating the outbreak, the once *pathogenic* infections may return to the *latent* condition in the surviving fish – waiting for the next downturn in the environment.

LEECH INFESTATION

Caused by
Piscicola geometra and various other leeches.

Obvious symptoms
Large leeches (up to 5cm/2in long) firmly attached to the skin, fins and, perhaps, gills. Heavily infested fish may appear listless, thin and occasionally behave in an agitated fashion. Reddened areas on the body indicate previous points of attachment, which may become infected with fungus. Leeches may, via their bloodsucking feeding habits, transmit microbial diseases between fish.

Occurrence of the problem
Leeches and other flatworms are often introduced into ponds and, more rarely, aquariums with new fish, plants (especially plants from local ponds or rivers) or live food. Not all flatworms are parasitic, however; some are simply scavengers.

Leeches reproduce by means of eggs, and these are laid in ponds in temperate regions anytime from spring to autumn. These eggs are quite resistant to treatment and, once they hatch out, the parasites can live for some time away from the fish host.

Treatment and control
Organophosphorus insecticides, such as metriphonate, may be used to treat leech infestations, although the active ingredients may not eliminate the eggs. Consequently, more than one treatment may be necessary. Note that some fish and many invertebrates are very sensitive to these chemicals. Leeches have also been removed from freshwater pond fish by placing the infested fish in a 2-3 percent solution of cooking salt (sodium chloride) for 15-30 minutes.

Leeches are most often a problem in ponds rather than in aquariums. One of the most effective ways of eradicating leeches from a pond is to drain and dry the pond for several days, discarding any plants and replacing them from a leech-free source. Adequate drying will kill both leeches and larvae within eggs.

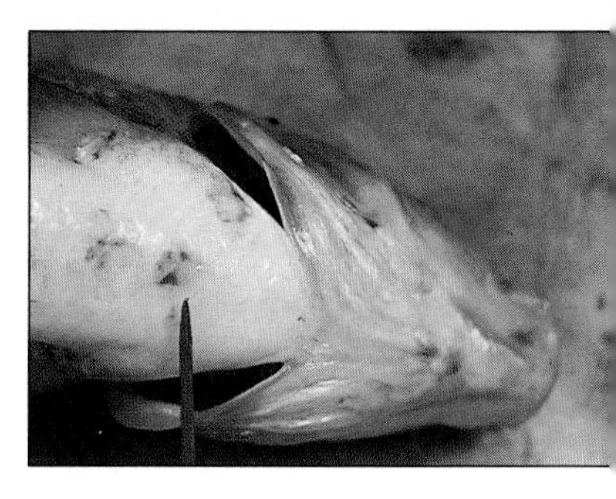

Above: Typical damage to the skin of a fish caused by the bloodsucking feeding habits of leeches. Such lesions may become infected.

Below: Goldfish may suffer from heavy infestations with leeches as they emerge from their winter hibernation in garden ponds.

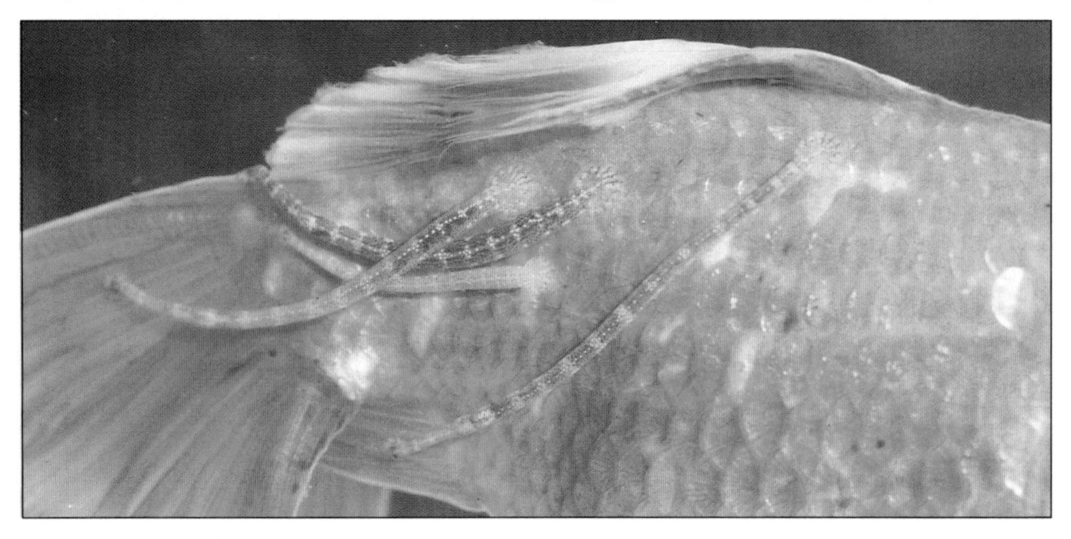

The life cycle of the fish leech (*Piscicola geometra*)

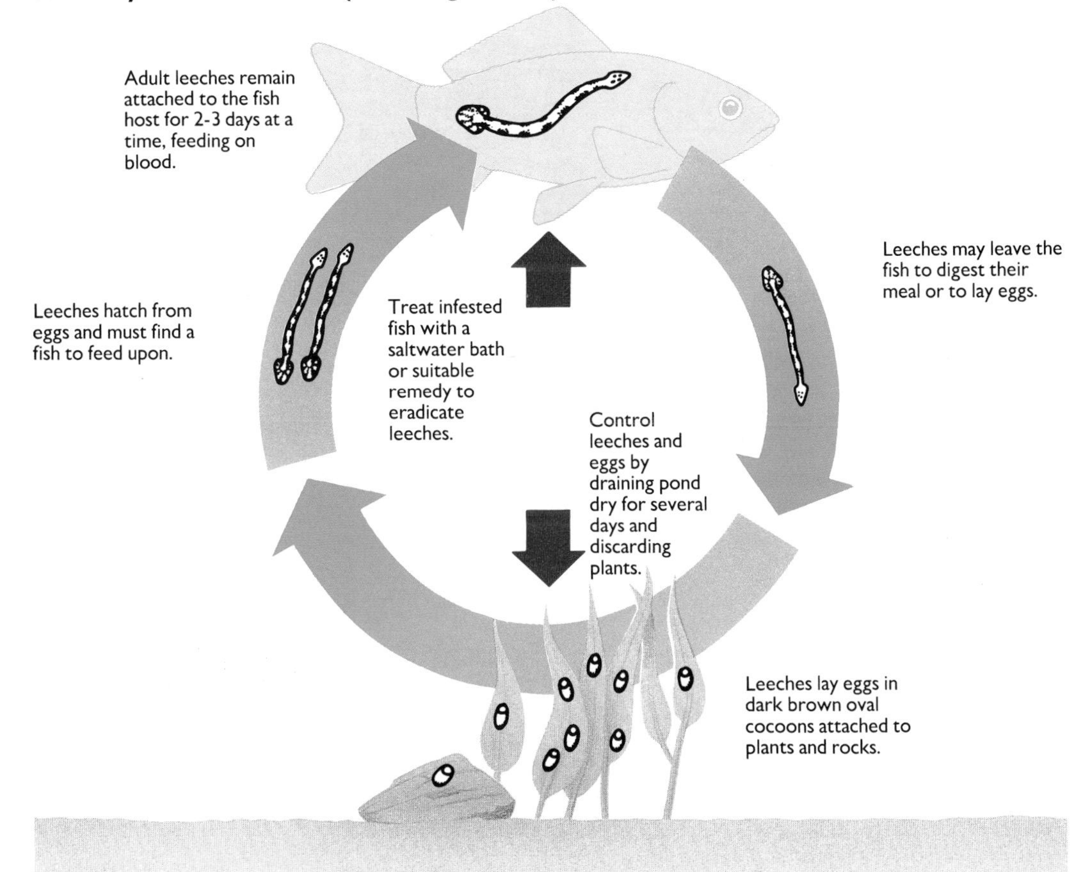

WHAT IS A LEECH?

Leeches are classified in the phylum Annelida, together with other 'worms', such as earthworms, bristleworms, whiteworms, and *Tubifex* worms, etc. As a group, the annelids usually have muscular, well-organized bodies, with a well-developed nervous system and alimentary tract. Most are free living.

The leeches are placed in a separate grouping within the annelids – the Hirudinea. There are over 300 species of leeches. Most live in fresh water; a few live on land or in the sea. They usually have a sucker at both ends of the body, and these suckers are powerful attachment organs. Most leeches measure a few centimetres in length; some are considerably longer.

Some leeches are predatory, feeding on small invertebrates, but many are parasites that feed on the blood and tissue fluids of other invertebrates, as well as fish, amphibians and other vertebrates. Bloodsucking leeches have sharp, cutting mouthparts, and secrete an anticoagulant to prevent the blood from clotting. A single leech can, in a single meal, take in ten times its own weight of blood and then fast for several months.

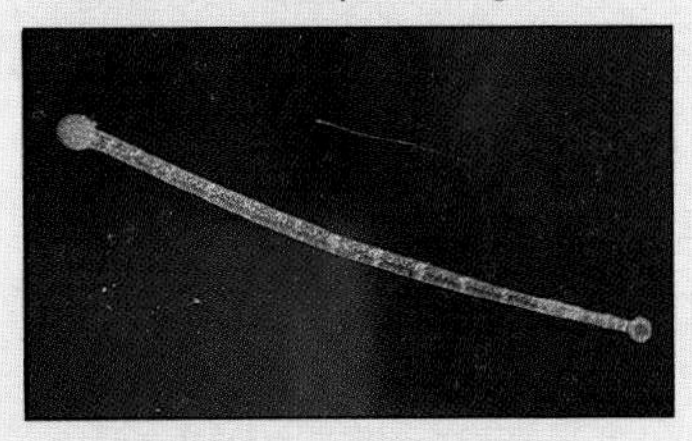

Above: A fish leech, *Piscicola geometra*, engorged with blood after feeding. Up to 5cm(2in).

Leeches are sometimes confused with turbellarian flatworms, although the latter are usually small, without obvious suckers, and move with a smooth gliding motion. Although leeches are hermaphrodite (i.e. contain male and female sexual organs), they cannot fertilize themselves, but need to cross-fertilize with another leech. Leeches may lay their eggs in cocoons above the water level. The eggs are resistant to many chemicals, but are usually susceptible to thorough drying.

NEON TETRA DISEASE

Caused by

Infection with the microsporidian parasite, *Pleistophora*. (For more details of these single-celled parasites, see the panel on page 169.)

Obvious symptoms

Fish may carry low-level infections without showing any ill-effects. However, heavily infected fish exhibit a loss of coloration (especially the red stripe in neon tetras), unusual swimming behaviour, spinal curvature, emaciation and finrot. A range of fish species are susceptible to this type of infection but many tetras seem particularly susceptible. The zebra danio and some barbs are also commonly affected by a similar disease.

Occurrence of the disease

The parasites pass from fish to fish, although the spores may be capable of living away from the fish host for a short while. Keeping fish under less than optimum conditions seems important in triggering the disease. Secondary bacterial infections are common.

Treatment and control

Although treatment has been attempted using a variety of chemicals, none has been completely effective. Experimental use of the anti-coccidian drug, toltrazuril (see page 196), has shown some success. Furazolidone (see page 188) appears effective in controlling secondary bacterial infections (so-called 'false neon disease'), rather than acting on *Pleistophora* directly. Always isolate fish showing symptoms of this disease from other stock, and attempt therapy. Even so, fish that recover following treatment may still harbour the parasite. Affected fish that do not respond should be painlessly destroyed, and it is vital to prevent cannibalism of dead or moribund fish.

Since apparently healthy fish may carry the infection, it is very important to maintain correct aquarium conditions in an effort to prevent the disease becoming a problem. Strip down any tanks which have suffered from a severe outbreak of the disease, thoroughly disinfect and rinse them and allow them to dry. The tank may then be refurbished with new decorations and restocked with fish from an alternative source.

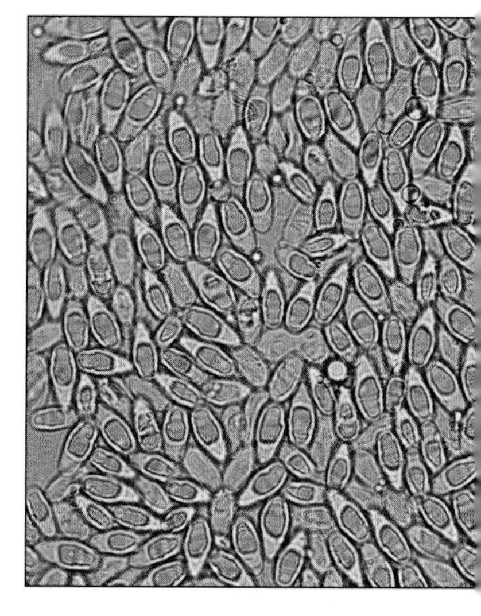

Above: This photograph taken down a high-power microscope shows spores from a *Pleistophora* cyst in infected tissue. One cyst contains thousands of spores, each about 5 microns across.

Below: Classic symptoms of neon tetra disease, notably the loss of colour on this normally vividly marked neon tetra. This fish is also emaciated and has finrot.

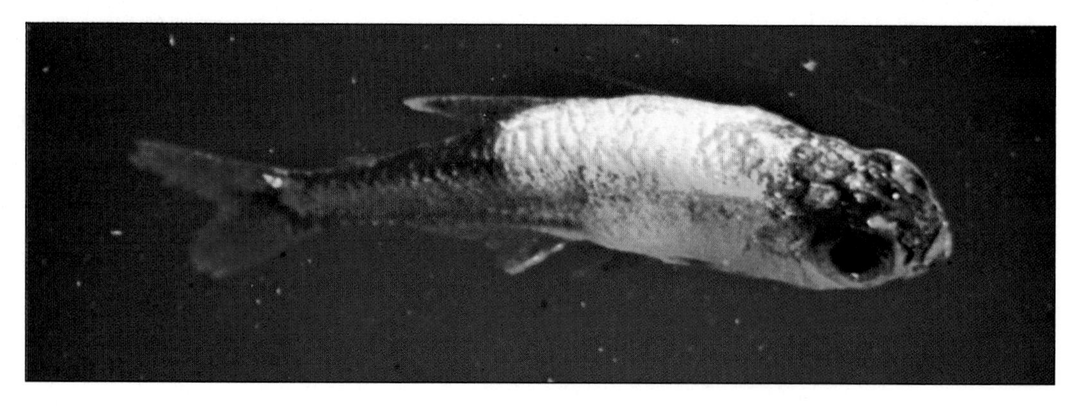

The life cycle of *Pleistophora*, the parasite responsible for neon tetra disease

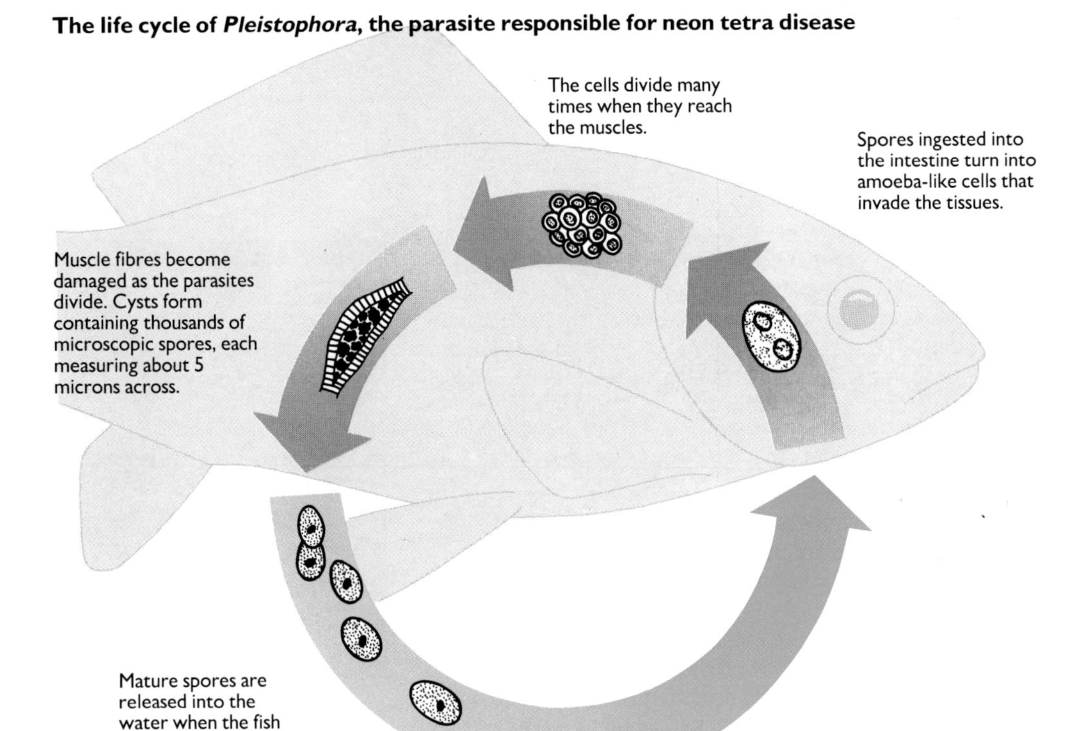

PARASITE LIFE CYCLES AND HOST SPECIFICITY

Some parasites have simple, or *direct*, life cycles, involving only one type of host. White spot, caused by the single-celled parasite *Ichthyophthirius*, is an excellent example of this kind of life cycle (see page 166). Other parasites have more complex, or *indirect*, life cycles, which may involve two or more very different types of host. Many of the internal helminth ('worm') parasites, such as *Camallanus*, have indirect life cycles (see page 176). The host in which the parasite matures is called the *definitive* host and the other hosts are usually referred to as *intermediate* hosts. The intermediate host can be very important in allowing parasite development and/or multiplication to occur within its tissues, and in aiding the transmission of the parasite to the definitive host. Without its intermediate host, the parasite cannot complete its life cycle.

Of course, parasites with direct life cycles can usually pass very easily between hosts in a heavily stocked aquarium or pond, which is why parasites such as white spot can have such devastating effects. By comparison, parasites with indirect life cycles rarely have all their necessary hosts within an aquarium or pond, and hence do not cause the same problems. Some parasites can infect a wide range of host species, and thus exhibit *low host specificity*. Parasites such as *Ichthyophthirius* (white spot), *Piscinoodinium* (velvet disease) and *Lernaea* (anchor worm) fall into this category. Parasites with direct life cycles *and* low host specificity are among the most troublesome to fishkeepers, because they can rapidly build up to large scale disease proportions and affect a range of fish species.

Other parasites can infect only a limited number of host species, whether these are definitive or intermediate hosts in the life cycle. Such parasites are described as having *high host specificity*. Highly host specific parasites are usually closely adapted to the conditions which occur on or in their preferred hosts. Even if such parasites can survive in or on other hosts, it is unlikely that they will be able to mature and reproduce there. Many of the intestinal digenetic fluke parasites of fish are quite host specific – the yellow grub, black spot and eye fluke parasites described on page 178, for example – as are some of the micro- and myxosporidian (sporozoan) parasites, such as neon tetra disease. Parasites with indirect life cycles *and* high host specificity do not often cause problems to fishkeepers, because they infect only a limited range of fish species and all the hosts required to complete their life cycles are rarely present.

NODULAR DISEASES

Caused by
Various microsporidian and myxosporidian parasites, such as *Ichthyosporidium, Nosema, Myxobolus* and *Henneguya*, and *Dermocystidium.*

Obvious symptoms
Small to large smooth, yellowish white cysts on skin, fins, gills, in muscle or among internal organs. Cysts may vary from a few millimetres up to a centimetre (0.4in) or so in size. They are usually spherical or oval, although some are elongate or irregular in shape. Each cyst contains many thousands of tiny parasitic spores. These spores are so small that it may take at least 15,000 to cover the head of a pin.

Occurrence of the disease
These parasites are similar to those causing neon tetra disease (see page 126). While low-level internal infestations probably occur unnoticed by the aquarist, infestations of the skin and fins are much more obvious. Although the spores are thought to pass from fish to fish, the life cycles of most of these parasites are poorly understood. Very heavy infestations, especially on the gills or on small fish, may be debilitating, although low-level infestations do not seem to do much harm. Each nodular disease may be specific to one or a small number of fish species and hence cross infections are unusual.

Treatment and control
There is no reliable treatment for this type of disease. Maintain infected fish in isolation from other fish; if their condition deteriorates, destroy them painlessly (see page 200). Disinfect tanks and all associated equipment which have contained infected stock as described on page 198. Do not buy obviously infected fish.

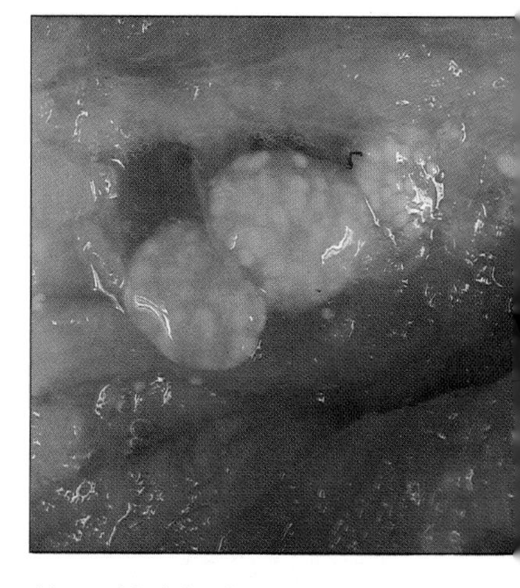

Above: Nodular disease cysts may affect the internal organs, here in the spinal cord of an angler fish, and cause swellings that are visible from outside.

Below: Such white or off-white cysts on the skin or fins, here on a three-spined stickleback, are a clear indication of an infestation with nodular disease.

The possible infection cycles of some nodular diseases of fish

Typical white or yellow cysts on or inside the body contain thousands of tiny spores, each measuring about 10 microns.

Spores released from infected fish may be eaten by other fish, thus passing on the infection.

Fish may pick up spores by eating infected fish.

Some nodular disease parasites may produce cysts inside the body of the fish.

Invertebrates may release spores or are eaten directly by fish.

Control spread by avoiding live foods that may be infected.

Spores released from fish or from the decomposed bodies of dead fish.

Invertebrate hosts, such as *Tubifex* worms, may be involved, allowing the spores of some species of parasites to 'mature'.

NUTRITIONAL PROBLEMS

Caused by
Incorrect diet.

Obvious symptoms
Symptoms may vary according to the nature of the problem. A diet that is deficient in protein, for example, may cause reduced growth rate and perhaps curvature of the spine. An excess of carbohydrate or fat (lipids) in the diet may produce a variety of symptoms, including liver disorders, anaemia and even an increased susceptibility to certain infectious diseases. Vitamin deficiencies may also occur in fish. A lack of vitamin A, for example, will induce poor growth, blindness and haemorrhaging at the base of the fins. The symptoms of vitamin B shortage include abnormal coloration and unusual swimming movements, with periods of extreme excitability interspersed with paralysis. Vitamin C deficiency may cause skin lesions and 'head and lateral line erosion' (HLLE), particularly in marine fish (see page 122). Little is known about the mineral requirements of fish, although a diet that contains the wrong amounts of calcium, magnesium and potassium may lead to kidney and intestinal problems. A further indication of the signs of vitamin deficiencies is provided in the table shown opposite, although it is important to note that this represents only a sample of recorded deficiency problems and that these symptoms can vary quite substantially between different fish species.

Above: Cloudy eyes, here in *Pangasius sutchi*, can be caused by nutritional problems, as well as by parasites and damage.

Below: Trout deformed through vitamin C deficiency, just one of the possible results of feeding a poor or incorrect diet.

Occurrence of the problem

Nutritional problems are rare in wild fish, and similarly unusual in pond and aquarium fish that are fed a varied diet based on good-quality prepared foods, such as flaked, tablet and pellet foods.

Treatment and control

Quite obviously, nutritional diseases can be avoided by feeding a good varied diet. 'Special' foods, such as vegetable-based freeze-dried and gamma-irradiated frozen foods, are available to supplement and add variety to a staple diet of flaked foods, for example, and 'safe' live foods can be used where relevant.

Always store opened pots of dry foods (e.g. flakes, pellets) under cool, dry conditions. This will help preserve their nutritional value.

Of course, it is important to avoid overfeeding with concentrated dried foods, since their accumulation in the pond and aquarium can have drastic and deleterious effects on water quality, and even bring about fish losses.

Some recorded vitamin deficiency problems of fish

Vitamin	Deficiency symptoms may include
Vitamin A	Poor growth, loss of appetite, eye problems, dropsy, gill problems, haemorrhaging at fin bases
Vitamin B	
Thiamine (B_1)	Poor appetite, muscular wasting, convulsions, loss of equilibrium, accumulation of fluid in cells (oedema), poor growth
Riboflavin (B_2)	Cloudy eyes, blood-shot eyes, poor vision, avoidance reaction to light, dark coloration, poor appetite, poor growth, anaemia
Pyridoxine (B_6)	Nervous disorders such as extreme irritability and fits, loss of appetite, anaemia, accumulation of fluid in the body cavity, gasping, flaring of gill covers
Cobalmine (B_{12})	Poor appetite, anaemia, poor growth
Un-numbered vitamins in the B complex	
Biotin	Loss of appetite, muscular wasting, convulsions, skin and gut lesions, poor growth
Choline	Poor growth, visceral haemorrhages
Folic acid	Poor growth, lethargy, fin damage, dark coloration, anaemia
Inositol	Poor growth, dropsy, skin lesions
Pantothenic acid	Gill and skin problems, loss of appetite, poor growth, lethargy
Vitamin C	Dark coloration, skin problems, eye diseases, spinal deformities

PESTS 1: ALGAL PROBLEMS

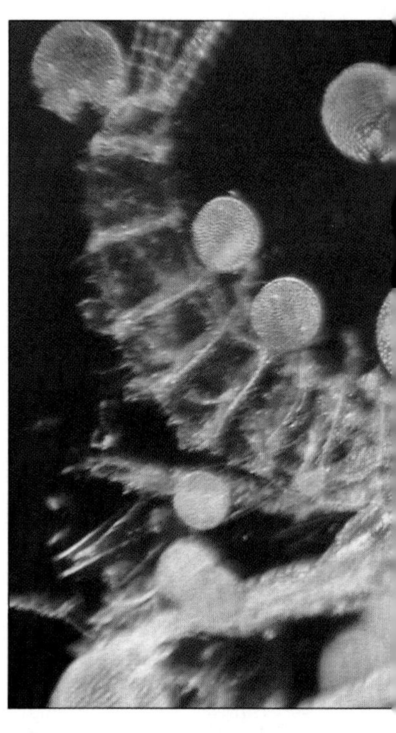

Caused by
A range of freshwater and marine algae, including diatoms.

Obvious signs
In freshwater, single-celled or colonial algae, such as *Chlamydomonas, Chlorella, Volvox* and *Scenedesmus,* give rise to familiar 'green water' conditions, while *Cladophora, Oedogonium, Vaucheria* and *Spirogyra* are among the filamentous types that form threadlike masses over plants, rocks and gravel, and include the well-known 'blanketweed'. Brown or yellow-brown filmlike growths in both fresh water and sea water are usually caused by various diatoms, single-celled algae with siliceous cells walls, that form part of the 'plant plankton' of open waters. Dark green (even black) slimy 'sheets' formed over rocks and gravel and/or an oily scum at the water surface are indications of the presence of blue-green algae (cyanobacteria), such as *Anabaena, Oscillatoria* and *Rivularia*. More often a problem in marine aquariums are various types of red algae that form films over rocks and gravel.

Together with the obvious presence of heavy growths of algae there may be certain water quality problems, such as high or fluctuating pH, periodic low dissolved oxygen levels and signs of gill irritation or even poisoning of the fish.

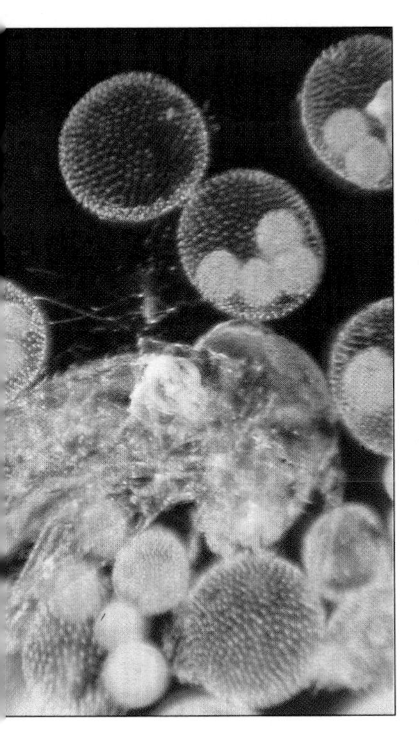

Above: Free-swimming colonial algae, such as these spherical *Volvox,* are one cause of green water in an aquarium or pond. Each colony is up to 1mm (0.04in) across.

Above right: Algae completely smothering an aquarium plant. Ideally, try to tackle such algal problems well before they reach this advanced stage.

Left: Filamentous green algae can coat aquarium plants, gravel and decorations, as shown here. This may result from too much light or too few plants in the tank.

Occurrence of the problem

A small amount of algae in an aquarium or pond is, of course, quite normal – even attractive – and probably beneficial to the fish. Similarly, new (and sometimes established) aquariums and ponds may experience temporary blooms, which soon subside of their own accord. What is of more concern are persistent or recurring algal blooms, and their effects on general water quality and fish health in the aquarium or garden pond. (See Chapters 3 and 4 for further information on maintaining water quality.)

A variety of causes have been put forward to explain certain types of algal blooms in ponds and aquariums. Excessive growth of green algae, either in the form of green water or as a filamentous growth, are often the result of too much light – especially natural sunlight or tungsten bulb lighting – and/or too few aquarium plants. Nutrient-rich garden run-off may also be important in causing algal blooms in ponds. Plants actively compete with, and may thus help to control, algae for available light energy and dissolved nutrients. The characteristic brown film-like growths of diatoms usually occur in relatively dim lighting conditions, and blue-green algae may be a sign of generally poor water quality (including high levels of some nutrients and infrequent partial water changes) and high light levels. The cause(s) of excessive growths of red algae in marine tanks are, perhaps, less well understood, although may include inadequate lighting, low temperatures and infrequent partial water changes.

Clearly the causes of algal blooms in ponds and aquariums are many and varied, but the occurrence of such a problem invariably indicates that something is wrong with the prevailing environmental conditions and general pond or aquarium care.

● Initially, remove as much of the offending algae as possible, using a scraper and siphon tube. Filamentous growths in a pond can sometimes be removed with a garden rake. If possible, remove rocks, decorations and plants for cleaning; if adequate cleaning is not possible, you may need to discard them. Hot water and a stiff brush will remove algae from most hard surfaces. Algae present as green water in a garden pond can be removed by a large-scale water change. However this often provides only a temporary solution for the algae may soon return. A more effective method is to fit a UV unit to the outflow pipe of the pond filter. The UV irradiation kills the algae cells as they pass through the unit, causing them to clump together (flocculate). Barley straw appears to have a controlling effect on blanket weed and pads of insecticide-free barley straw are now commercially available for use in ponds.

● Having removed much of the algae, it is usually possible to treat green filamentous algae and green water safely and effectively using a proprietary algicide. Be sure to follow the manufacturer's instructions to ensure safe and efficient use.

● Investigate the tank or pond conditions:
Are there enough plants? An aquarium 1m (39in) long will require at least several dozen plants at the recommended lighting levels. If it receives substantial amounts of sunlight, a garden pond should have up to 60 percent of its surface area shaded by lily leaves (or similar). Plants and shading will help to reduce the amount of nutrients and light available to the algae.

Is there too much light or too little light? Excessive amounts of light – particularly natural sunlight and tungsten bulb lighting – especially in the absence of significant numbers of plants will encourage some algal problems. Dim light conditions, on the other hand, may encourage diatoms and red algae.

What are water conditions like? A build-up of dissolved nutrients, infrequent partial water changes and perhaps other related factors may precipitate algal blooms.

● Fish and some invertebrates can be used to help control algae. In tropical freshwater aquariums, fish such as mollies (*Poecilia* spp.), some *Labeo* sharks, the sucking loach (*Gyrinocheilus aymonieri*), and certain sucker-mouth catfish (e.g. *Otocinclus*) will consume algae. The common plec catfish (*Hypostomus plecostomus*) is an efficient algae-eater but has the disadvantage of reaching a large size, often exceeding 0.3m (12 in). Freshwater algae-eating

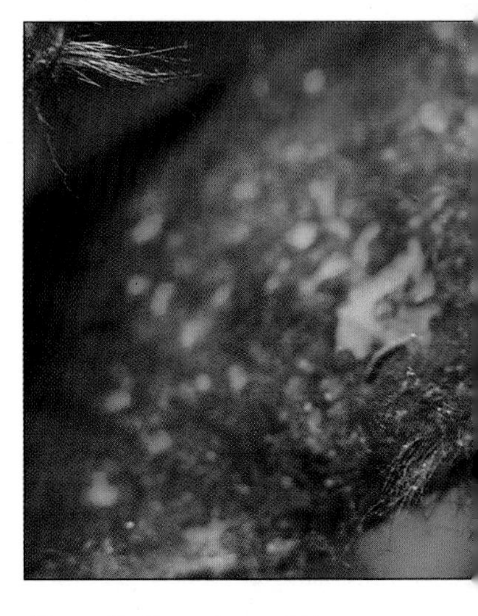

Above: Snails can be useful in controlling unwanted algal growth, simply by eating the excess, but they may multiply and become a problem themselves.

Below: Certain catfish, such as *Otocinclus* (as shown here), are very efficient browsers on algae – so efficient, in fact, that they may need to have their diet supplemented with vegetable flaked foods.

shrimps are also available. For ponds, the grass carp (*Ctenopharyngodon idella*) is sometimes sold for controlling algae, but this relatively unattractive fish grows very large and tends to consume aquatic plants in preference to algal pests. Certain marine fish, especially surgeon fish (Acanthuridae) and dwarf angelfish (*Centropyge*), and some marine invertebrates, such as sea urchins (*Diadema*), enjoy browsing on green algae.

● As a last resort, proprietary algicides are available for treating green filamentous algae and green water in aquariums and ponds. Be sure to follow the manufacturer's instructions to ensure safe and efficient use.

It is important to note that algal blooms can suddenly die back of their own accord when, for example, the nutrient supply is exhausted. Such natural die-backs, along with algicide-induced die-backs, can produce large amounts of dead algae, with disastrous effects on water quality.

WHAT ARE ALGAE?

Algae (singular 'alga') are primitive plants, quite closely related to bacteria. They range in size from single cells ('micro-algae'), which are individually invisible to the naked eye, to the familiar brown seaweeds ('macro-algae'), which may grow over 50m(164ft) long. Approximately 25,000 species of algae are known.

While some algae are just single cells, others form filamentous masses, or even complex, many-celled 'colonies'. However, algae never have the well-defined roots, stems and leaves seen in higher plants.

Algae consist of five main groups:

● **Green algae** (Chlorophyta) – which include the single-celled or colonial algae responsible for 'green water', as well as many filamentous green algae and green seaweeds, such as *Ulva*.

● **Yellow-green algae and diatoms** (Chrysophyta) – which include *Vaucheria* (the common cause of dark, hairy growths over aquarium plants), and diatoms.

● **Blue-green algae**
These are not true algae but a group of primitive micro-organisms known as cyanobacteria. In the wild, some cyanobacteria are capable of producing toxins that are poisonous to fish.

● **Red algae** (Rhodophyta) – a group of principally marine algae, which range in form from small single-celled or filamentous kinds to branching or sheetlike seaweeds.

● **Brown algae** (Phaeophyta) – this includes the larger seaweeds, such as kelp and wracks. They are principally marine, and especially common in cooler regions.

All algae contain the pigment chlorophyll, although in some forms this green pigment may be masked by brown, red or yellow pigments. Chlorophyll allows algae and other plants to convert carbon dioxide and water (in the presence of bright light) into carbohydrates and oxygen in the process known as 'photosynthesis'. This use of carbon dioxide by plants (and algae) can give rise to quite marked diurnal fluctuations in pH and oxygen levels.

Reproduction in algae ranges from simple cell division to complex sexual processes. Algal spores may be carried many miles in air currents.

As a group, the algae are very important primary food producers and are fed upon by a huge range of animals. However, in the aquarium or pond they can sometimes get out of control.

Above: Some species of marine algae are quite attractive and should be encouraged in the marine aquarium. This red form is among the more unusual ones.

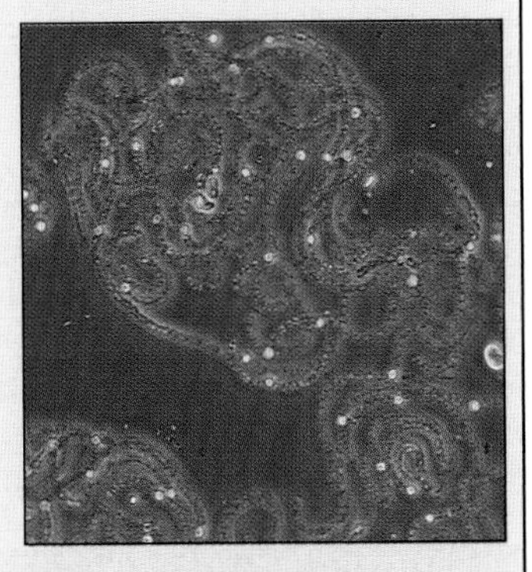

Above: *Anabaena*, a blue-green alga that commonly causes algal blooms in garden ponds. Strands are up to 30 microns across.

PESTS 2: FLATWORMS, BRISTLEWORMS, OSTRACODS, COPEPODS AND MITES

Caused by
A range of free-living animals, such as planarian flatworms, annelid bristleworms, ostracods, copepods and mites.

Obvious signs
Such animals appear to suddenly 'bloom' in an aquarium. Small flatworms, which can be white, cream, red or orange in colour, usually move with a smooth gliding motion. They are usually a few millimetres to, perhaps, a centimetre in length (up to 0.4in). Bristleworms are usually a little longer than this and often appear as pale tubificid-type worms among the gravel. Ostracods are small (to 3mm; 0.1in) bean-shaped crustaceans, typically yellow or brown-black in colour, that scuttle over aquarium surfaces. Tiny copepods, only just visible to the naked eye, look like clouds of moving detritus when present in large numbers, and tiny mites (less than a millimetre in size) may appear from time to time, usually in the damp area between the water surface and the cover glass.

Occurrence of the problem
Small numbers of these animals probably do little harm and, even when present in large numbers, they seem to be unsightly rather than dangerous. They are probably introduced with live food, living rock and live plants, etc. These pests appear to occur more extensively in tanks which have large amounts of organic matter, especially in the gravel, although mites may occur independently of this. They frequently 'bloom' then disappear of their own accord.

Above: Flatworms on the front glass of a freshwater tropical aquarium. They have arrow-shaped 'heads' and move with a gliding motion. About 1cm(0.4in) long.

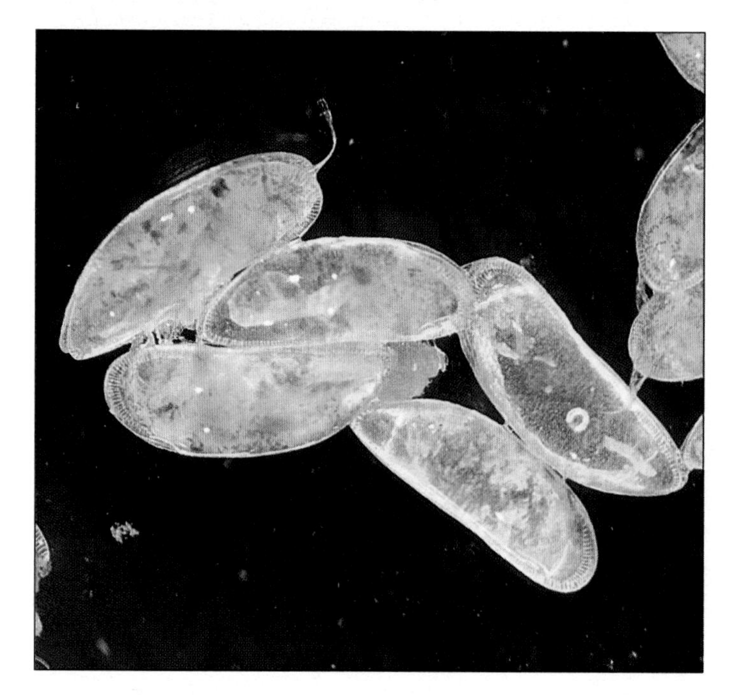

Left: Bean shaped ostracods, as viewed under the microscope. Some species are active swimmers.

Right: Mites often have tiny hairs over their bodies, which can produce a skin rash in some people. These tiny creatures are no more than 1mm(0.04in) long.

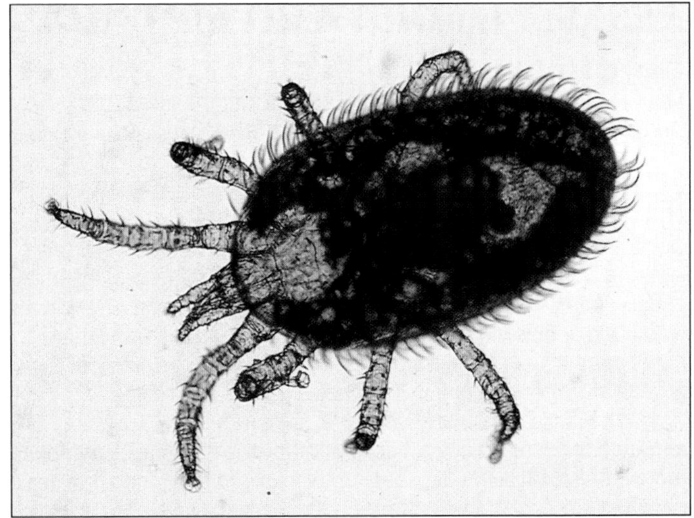

Treatment and control

From what has been said, it is clear that avoiding overfeeding, regular 'hoovering' of the tank substrate and routine filter maintenance will help prevent and perhaps control these pests. In addition, fish such as *Betta* species, some gouramis and kribensis (*Pelvicachromis pulcher*) will feed on flatworms. Raising the tank temperature to 35°C(95°F) for several hours, after removing all the fish, is also said to kill flatworms (see also *Hydra*, page 138). Mites can be removed from around the top of the tank using a damp cloth.

Although these pests rarely cause problems to warrant such a drastic approach, a certain cure is to completely strip the tank down, rinse all the decorations and equipment in dilute bleach (and then in running water), before refurbishing the tank with new plants and clean water. Preventing these animals from being re-introduced at some time in the future will, however, prove difficult, and hence careful management is the best approach to the problem.

WHAT IS A FLATWORM?

The flatworms, or platyhelminths, are all characterized by their flattened, often leaflike shape. They are divided into four main groups:

- **Turbellaria** – usually free-living flatworms, some of which are also known as planarians.
- **Monogenea** – parasitic monogenetic (one host) flukes.
- **Digenea** – parasitic digenetic (two or more host) flukes.
- **Cestoda** – parasitic cestodes, or tapeworms.

Here, we consider the free-living turbellarians; the parasitic platyhelminths are featured on pages 170-181.

There are 1600 species of turbellarians. They are aquatic and occur in both fresh and salt water. They can often be found beneath stones in ponds, streams and on the seashore.

Turbellarians are usually about 1cm(0.4in) or so in length. Their ciliated body surface often gives them a smooth, gliding motion. These flatworms are generally carnivorous and feed on all kinds of small live or dead insects and crustaceans.

In common with most other platyhelminths, turbellarians are hermaphrodite (i.e. each individual contains both male and female reproductive organs) and are able to reproduce sexually, usually laying eggs in small cocoons. They also have excellent powers of regeneration. If cut into several pieces, for example, each will grow into a complete, new animal. *Dendrocoelum* and *Polycelis* are turbellarians often used in practical classes in schools and colleges, and at least one turbellarian has now been implicated in a skin disease of marine fish (see *Sliminess of the skin*, page 146).

PESTS 3: HYDRA AND AIPTASIA

Caused by

The freshwater coelenterate *Hydra* and the marine anemone *Aiptasia*.

Obvious signs

Hydra have a small, stalklike body, which ends in up to ten long, slender tentacles. Clustered on these tentacles are batteries of tiny stinging cells, which *Hydra* uses to capture and kill its prey of small crustaceans, other invertebrates, and even small fishes. *Hydra* may reach two or more centimetres (approximately an inch) in length, but can rapidly contract its body, so that it appears as no more than a small piece of rubbery jelly attached to plants and rocks.

Aiptasia is in some ways similar in appearance to *Hydra*, although it can reach at least several centimetres in length and has more tentacles. The tentacles are of two sizes: a few long and very obvious ones, and a great many smaller, less obvious ones. Like *Hydra*, the tentacles contain stinging cells used in prey capture, and also like *Hydra*, this anemone can contract quite suddenly and markedly when disturbed. Although it does not seem to harm fish in a marine tank, it may have some adverse effects on nearby corals, if present in large numbers.

Above: A variety of three-spot gourami (*Trichogaster trichopterus*), one of several fishes that will eat *Hydra* in freshwater aquariums – an ideal form of biological control.

Below: A mass of *Aiptasia*. Common in warm and temperate seas, some species can assume pest proportions in the tropical marine invertebrate aquarium.

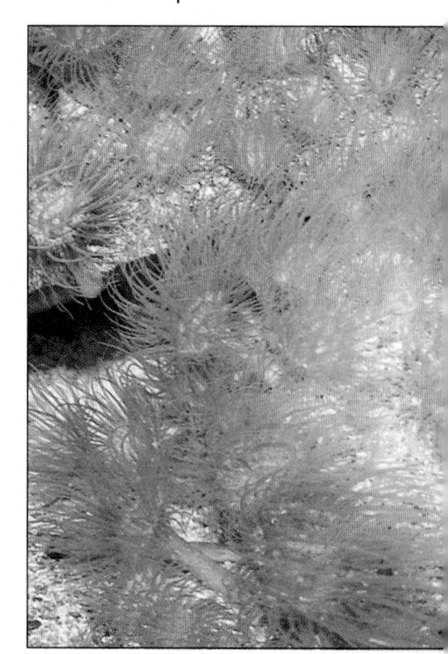

Occurrence of the problem

Hydra are usually introduced into a freshwater aquarium or pond with plants or live food. They may then multiply rapidly, until they can be seen in large numbers coating all the available submerged surfaces. *Hydra* is most often a problem in small aquariums, where they may kill and eat fish fry and even small fish that come too close to their stinging tentacles.

Aiptasia are usually introduced into marine tanks with 'living rock' (corals), and although these hardy anemones do little harm in, a fish-only system, they can build up to large numbers and compete with and overgrow delicate corals in an invertebrate system.

Treatment and control

Improving general hygiene, avoiding overfeeding and carrying out regular tank maintenance may moderate the build-up of these pests. When present in very small numbers, they can probably be controlled by removing infested items of tank decor and vigorously cleaning them in running water. However, more positive action will be required to control large numbers of these pests.

In freshwater aquariums, fish such as the three-spot gourami (*Trichogaster trichopterus*) and the paradisefish (*Macropodus opercularis*) will consume large numbers of *Hydra*. In a marine tank, some butterflyfish, such as the four-eyed butterflyfish (*Chaetodon capistratus*), will often consume *Aiptasia* without harming more delicate corals.

Hydra can also be eliminated by removing all the fish from the aquarium and raising the temperature of the water to around 40°C (104°F) for several hours. Most plants will tolerate this. Afterwards, carry out a substantial water change and carefully 'hoover' the tank floor with a siphon tube before topping up the

tank with conditioned tapwater at the correct temperature. Then reintroduce the fish.

Hydra is also quite sensitive to common salt, and exposure to a salinity of 0.3-0.5 percent for five to seven days will usually control the problem. Even many 'softwater' fish will tolerate this level of salt for a few days, but be sure to make a substantial water change and a return to more 'normal' water conditions after this period.

Aiptasia can be controlled by removing the worst affected pieces of decor and soaking them in fresh water for several days, although this will also kill any other resident marine organisms. Then rinse the rock or coral in running water, before returning it to the aquarium. As an alternative, use a small syringe fitted with a fine needle to inject one or two drops of a strongly alkaline solution of sodium hydroxide (NaOH) into the base of the stalk of *Aiptasia*. If done carefully, this will kill the pest and not harm surrounding marine organisms or affect the water chemistry.

Since effective control of both *Hydra* and *Aiptasia* can be problematical, it is vital to avoid introducing these pests accidentally. Avoid live foods from natural freshwater ponds and streams, and inspect new rocks, corals and plants, etc. closely (while immersed in a small amount of water from the aquarium or pond) before introducing them to your set-up system.

WHAT IS A COELENTERATE?

Coelenterates (phylum Coelenterata or Cnidaria) are a group of some 9000 species of principally marine invertebrates, with one or two examples in fresh water. The coelenterates consist of three main groups:

- **Hydrozoa**, which include the colonial marine millepore corals and freshwater *Hydra*.
- **Scyphozoa**, which includes the familiar jellyfishes.
- **Anthozoa**, which includes the sea anemones and the hard and soft colonial marine corals.

These simple, many-celled animals can exist in two forms: anchored as a 'polyp' (e.g. *Hydra*), or as free-swimming 'medusae' (e.g. jellyfish). Most have a mouth surrounded by a ring of stinging tentacles, used in defence and/or prey capture.

Reproduction is often quite complicated, and coelenterates can alternate between asexual and sexual phases. During the sexual phase, ciliated larvae called 'planula' are released, and eventually settle on a rock to give rise to a typical polyp. Polyps can give rise to more polyps by asexual budding, or produce sexual forms, which in turn give rise to more planulae larvae. These larvae are important for dispersing the coelenterates by means of ocean currents.

It is interesting to note that some coelenterates, such as *Hydra*, have quite remarkable powers of regeneration. If cut into pieces, each piece can regenerate into a new individual.

While the tentacles of some jellyfish, tropical corals and tropical anemones can be very painful to humans, smaller coelenterates, such as *Hydra* and *Aiptasia*, do not pose a threat.

Below: The elegant polyps of *Hydra*, a freshwater relative of the marine anemones and corals.

PESTS 4: SNAILS, BEETLES AND DRAGONFLIES

Caused by

Various species of freshwater snails, beetles and the larvae of some dragonflies. The water boatman (e.g. *Notonecta*) is another predatory aquatic insect.

Obvious signs

Large numbers of snails may build up, some species of which are active mainly at night, perhaps with coincident damage to plants. The adults and larvae of some freshwater beetles can kill fish as large as themselves. They feed by sucking the juices from their victims, leaving the dead fish with no obvious signs save for small puncture wounds. The larvae of some dragonflies also prey on fish and fish fry in fresh water.

Occurrence of the problem

Snails are by no means essential for the successful maintenance of an aquarium or garden pond, and some species, especially the familiar pond snail *Lymnaea* and the cone-shaped Malayan livebearing snail *Melanoides*, may rapidly build up in numbers. Snails, like beetles and dragonflies, may be introduced with live foods and/or plants, although the winged adult dragonfly may deposit eggs at the water's surface of a pond.

Left: At about 4cm(1.6in) long, this fearsome-looking dragonfly nymph (*Aeshna* sp.) is a keen-eyed predator and can pose a threat to small fish fry that venture near.

Right: An adult great diving beetle (*Dytiscus marginalis*), about 3cm(1.2in) long. Only such relatively large beetles are a hazard to pond or aquarium fish.

Below right: The curved jaws of the great diving beetle larva can pierce prey up to its own length of 5cm(2in). Digestive juices are pumped into the prey through canals in the jaws and the larva then sucks its digested victim dry and discards the husk.

Right: A freshwater pufferfish (*Tetraodon* sp.) investigating a snail in the aquarium. These inflatable fishes have powerful jaws to cope with the shell.

Below: These delicate and colourful ram's horn snails (*Planorbis* sp.) are less of a problem than many other types of snails and are, in fact, quite attractive in an aquarium.

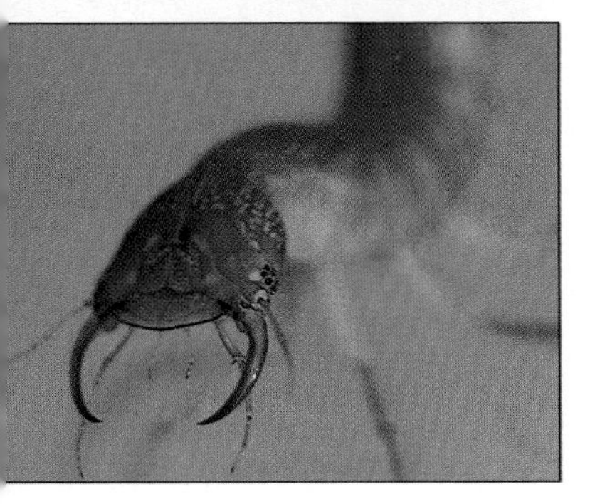

Treatment and control

Beetles, dragonflies and water boatmen are recorded as pests relatively infrequently in aquariums and ponds. Should a genuine problem develop from these animals, thorough cleaning and complete refurbishment is the only real answer. Rinsing all new plants in running water and avoiding suspect live foods will help to prevent introducing these insects.

Snails are recorded in large numbers in aquariums from time to time, although they are usually unsightly rather than a problem to fish or plants. Control methods include introducing fish such as the clown loach (*Botia macracantha*), convict cichlids (*Cichlasoma nigrofasciatum*) or pufferfish (*Tetraodon* spp.), which will feed on some snails. Alternatively, leave one or two fish food tablets on an upturned saucer on the tank floor overnight. The following morning, you can remove the saucer and the accumulated snails together. You will need to repeat this a number of times to have any real effect, and take care that uneaten food tablets do not pollute the aquarium. Chemical snail eradicators exist, although be sure to use these carefully, in case a sudden large die-off of snails causes a water quality problem. Should a snail problem warrant such an approach, empty the tank, clean everything thoroughly in dilute bleach and refurbish the tank using new plants and clean water. Be sure to remove all traces of bleach by repeated rinsing in running water.

As described on pages 98 and 180, snails can act as the intermediate hosts of a number of fish parasites. However, within the simplified conditions of an aquarium or pond, all the hosts necessary for the completion of such life cycles are seldom present. Do note, however, that some snails from tropical regions may harbour a range of parasites that can infect fish *and* humans, and so be sure to handle newly imported material very carefully.

PHYSICAL DAMAGE

Caused by

Physical damage can result from many causes, including rough handling, fighting, recent importation and parasite infestations.

Obvious signs

The most obvious signs are loss of scales and normal coloration, split or ragged fins, and damage to the body or mouth. Badly damaged fish may become listless, lying in a secluded corner of the tank or pond. If left untreated, any wounds may become infected with parasites, fungi or bacteria.

Treatment and control

It is vital to identify and eliminate the possible causes of damage. This may include handling fish more carefully – do not allow fin rays, for example, to become entangled in nets – separating aggressive fish or providing more refuges in the tank.

Treat slightly damaged freshwater aquarium fish in the set-up tank with a suitable antibacterial. Isolate pond fish and badly affected aquarium fish in a separate tank and treat them with a proprietary antibacterial plus a compatible tapwater conditioner or an antibacterial plus a small amount of salt (see Chapter 7).

It is best to isolate marine fish in a separate tank and treat them with a suitable aquarium antibacterial remedy. Keep a close watch for the development of further symptoms. These may indicate a more systemic bacterial problem or parasite infestation, which may require alternative treatment, perhaps with antibiotics.

To treat discrete, deep lesions, hold the fish carefully in a soft damp cloth and gently dab the affected areas with a suitable antiseptic, such as mercurochrome (see page 186).

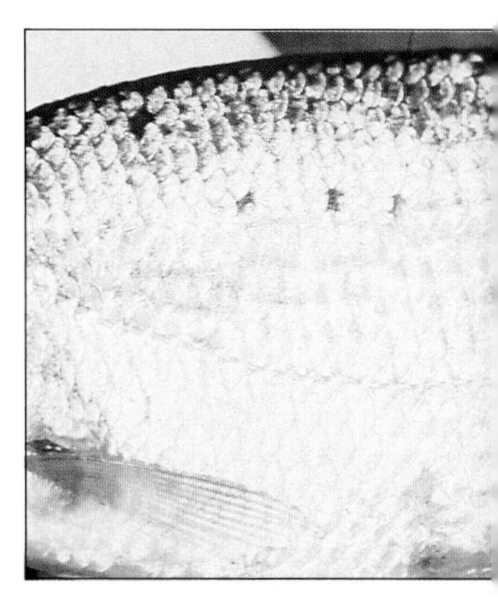

Below: Fighting and attempted cannibalism by other fish can often result in the loss of an eye, as in this cardinal tetra. Eyes are an easy and common target for such attacks.

Right: Fish such as piranhas (*Serrasalmus* sp.) can take bites out of each other, especially at feeding time. The wound on this piranha healed without treatment.

Below: Some lesions, such as this, resulting from rough handling and scale loss can quickly become secondarily infected with fungi and bacteria. Give prompt treatment.

Right: Damaged fins may become infected and develop finrot, as on the anal fin of this goldfish. When netting a fish, always use soft knotless netting.

HYGIENE AND HANDLING

During an outbreak of disease, the importance of pond and aquarium hygiene becomes paramount. Diseased fish can be a source of large numbers of infective organisms and, unless treatment is taking place *in* the aquarium or pond, it is important to remove such fish to an isolation tank. Of course, it is vital to remove dead or moribund fish promptly, since many species of fish are cannibalistic under such circumstances, which greatly aids disease transmission.

One sensible hygiene precaution is to avoid using the same equipment in diseased and apparently healthy ponds or aquariums. Disinfectants are available for cleaning contaminated equipment, although rinsing in running water and thorough drying does go some way towards eliminating many disease organisms.

The skin of fish is very delicate and easily damaged; even slight injuries can lead to the development of localized and eventually systemic infections. Try to prevent such probems by using soft 'knotless' netting in hand nets and avoiding any sharp items of equipment or decor. Fish are often best gently coaxed into a glass jar or shallow bowl, and then lifted, complete with water, from the aquarium or pond. Do not handle fish with bare hands. If you must handle them at all, use a soft damp cloth.

Personal hygiene is also important in fishkeeping, particularly when dealing with conditions such as wasting disease (fish TB) that can cause human infections. Be sure to wash your hands thoroughly after carrying out routine maintenance and do not immerse your hands in suspect water if you have uncovered cuts or abrasions.

POP-EYE OR EXOPHTHALMIA

Caused by

A variety of factors may be involved, including bacterial infection, parasite infestation, poor water quality, and internal (metabolic) disorders.

Obvious symptoms

One or both eyes protrude from the head in an unusual fashion. Note that some fish, such as moors and telescope-eyed goldfish, have been bred specifically for their 'pop-eyed' appearance.

Occurrence of the disease

This disease usually only affects one or two fish in a pond or aquarium and is rarely markedly infectious. Very often the condition persists for a short time and may then disappear. It is seldom fatal unless accompanied by dropsy (see page 104). If large numbers of fish suddenly begin showing symptoms of pop-eye then you should suspect the water quality or perhaps the presence of an infectious agent.

Treatment and control

It may be necessary to isolate and treat affected fish with a broad-spectrum antibiotic (preferably by injection), although this will only be effective if the symptoms are related to a bacterial infection. As an alternative approach, and bearing in mind the usual low infectivity of pop-eye, it may be best to leave affected fish in the set-up aquarium or pond and to give them optimum conditions and a good, varied diet. Carefully monitor their general condition and, if the disease appears to be spreading to previously uninfected stock, isolate the diseased fish from the rest. Should the affected fish show any signs of distress, destroy them painlessly.

Below: The pop-eyed appearance of this carp is not normal, although some forms of fancy goldfish are bred for their swollen eyes. In severe cases of pop-eye, the eye may burst from the socket and, although the wound will heal, the fish must cope with one eye.

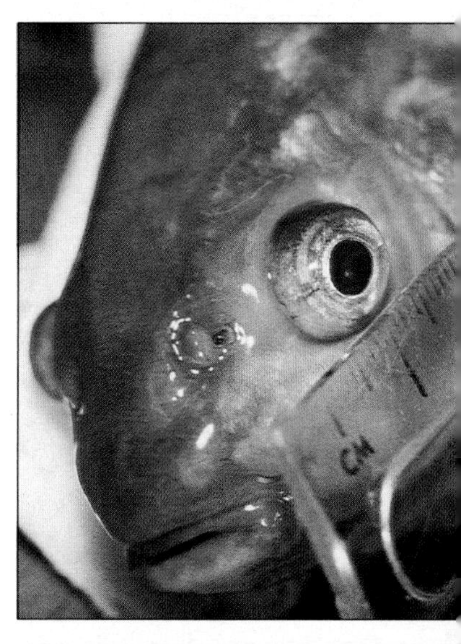

Above: A clear case of 'one-sided' pop-eye, which may or may not spread to the other eye. The causes of this condition appear to be many and varied, with no completely reliable means of treatment available.

Left: Pop-eye can be caused by a build-up of fluid in or behind the eyes or by the presence of bacteria and parasitic eye flukes. This oscar has a TB infection behind the eye, which is causing the swelling. The infection may have arisen from goldfishes already infected with TB given to the oscar as food.

Right: Both eyes are clearly affected in this fish. If many of your fishes show these symptoms, then check your water quality by using the simple test kits available to record the pH and nitrite levels particularly. (See Chapter 3, starting on page 32, for full details of using such test kits.)

SLIMINESS OF THE SKIN

Caused by

In freshwater fish various external protozoan parasites (such as *Ichthyobodo/Costia, Trichodina* and *Chilodonella*) and monogenetic flukes (such as *Gyrodactylus*) are involved. Infections with white spot (*Ichthyophthirius*) may also be present. In marine systems, the protozoan *Brooklynella*, the fluke *Benedina* and marine white spot (*Cryptocaryon*) may be involved. More recently, a turbellarian flatworm has also been implicated in this disease in marine fish.

Obvious symptoms

A grey-white film of excess mucus develops over the body and is especially noticeable over the eyes or areas of darkened pigmentation on the skin. Reddened areas may occur on the flanks of the fish, perhaps together with swollen gills and rapid movements. Badly affected fish become listless and lie on the bottom, occasionally scratching themselves against rocks. Secondary bacterial infection is common. A marine turbellarian infestation may also appear as small black spots (0.1mm/0.004in across) on pale fish.

Right: This microscope view reveals the ciliate protozoan *Chilodonella*, which is a common cause of sliminess of the skin in freshwater fish. These protozoans are about 50 microns across.

Below: A grey film over some or all of a fish, as shown here, plus scratching or rapid gill movements, are signs of sliminess of the skin.

Occurrence of the disease

This condition often occurs in fish that are in poor condition, perhaps because of overcrowding, poor water quality or inadequate nutrition. It readily passes from fish to fish and particularly affects coldwater fish in the spring as temperatures start to rise.

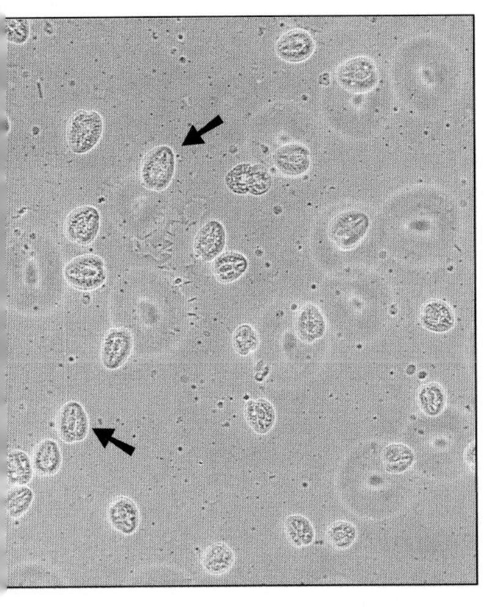

Right: The white film on the dorsal surface of this marine spiny boxfish indicates a possible infection with *Brooklynella*, the marine equivalent of the protozoan *Chilodonella*.

Below: It is rare for sliminess of the skin to be caused by just one kind of skin parasite. The excess mucus on this banjo catfish results from *Chilodonella* and the flagellate *Ichthyobodo (Costia)*.

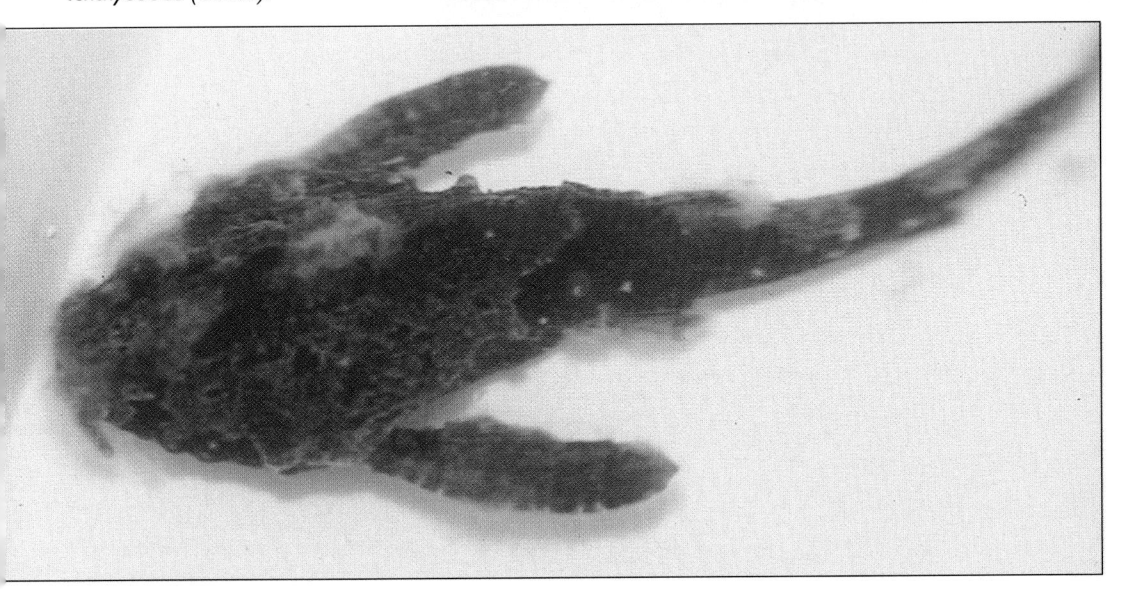

Treatment and control

Initially, treat affected freshwater fishes with a course of a reliable white spot remedy. Should this fail to bring about an improvement within five to seven days, carry out a 50 percent water change and use an antiparasite treatment based on formalin or an organophosphorus insecticide, such as metriphonate. Unfortunately, certain fish, such as orfe, rudd and piranhas, are rather sensitive to the latter (see Chapter 7).

For marine fishes, a short freshwater bath, followed by a course of copper treatment, may bring this problem under control. However, for stubborn outbreaks, try several one-hour formalin baths or use an organophosphorus insecticide. Formalin is particularly unpleasant; only use it in a treatment tank. Invertebrates are very intolerant of copper and insecticide-based remedies.

Eliminating the predisposing factors and quarantining all new fish are vital measures in the long-term control of this condition.

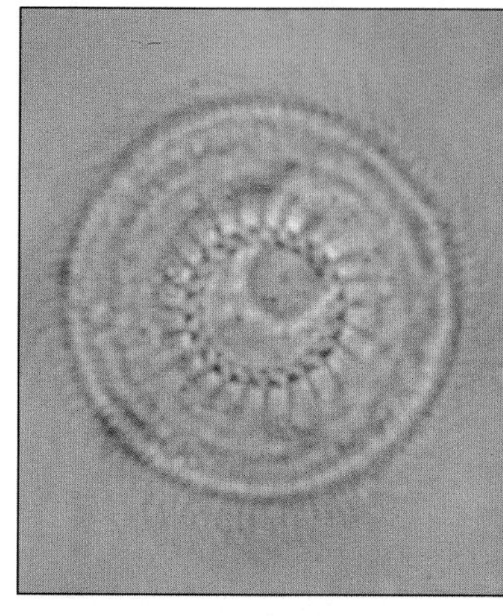

Above: The parasitic *Trichodina* protozoans are quite beautiful when viewed at x100-200 through a microscope. Variable in size, they are about 50 microns across.

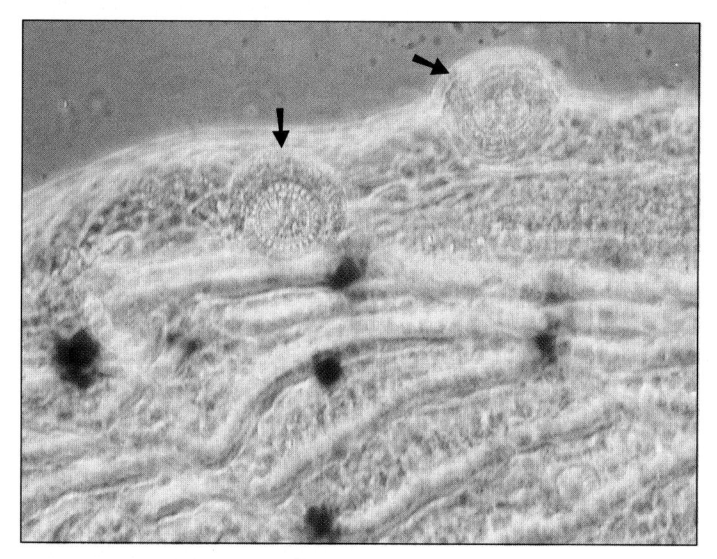

Left: A skin smear from a fish infested with trichodinid protozoan parasites, seen down a low-power microscope. Note the edge of the scale from the fish.

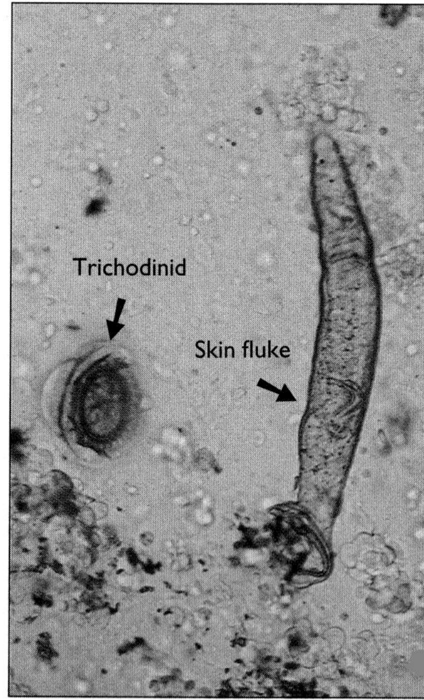

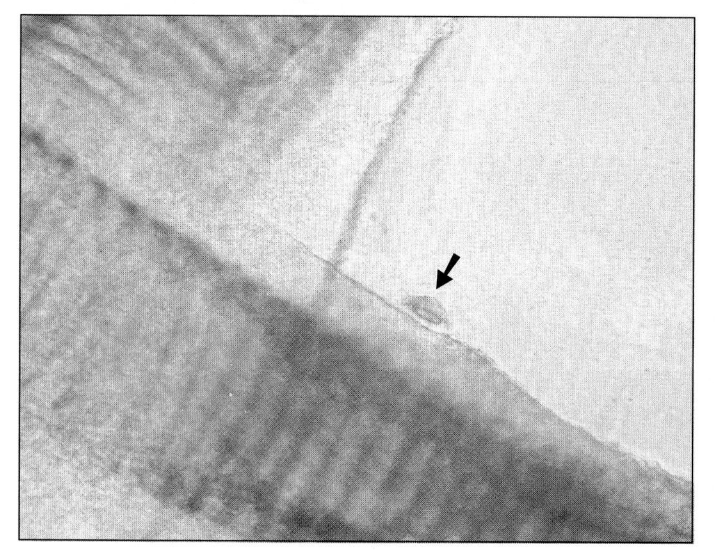

Left: Many of the parasites that cause sliminess of the skin can also cause gill problems. This trichodinid protozoan is shown edge on adjacent to a gill filament.

The life cycle of the skin fluke (*Gyrodactylus*)

Adult flukes live as parasites on fish skin. Their lifespan on the fish is 12-15 days at 15-20°C(59-68°F).

The hermaphrodite adults produce live young one at a time.

Within one day of birth these young themselves produce live young and then one every 5-10 days. The first young is born so soon because it started developing when its parent was inside its grandparent. In 30 days, a single *Gyrodactylus* can give rise to over 2000 more.

Suitable remedies will eradicate flukes on the fish and in the water.

Newly born flukes must find a fish host within 48 hours of birth to survive.

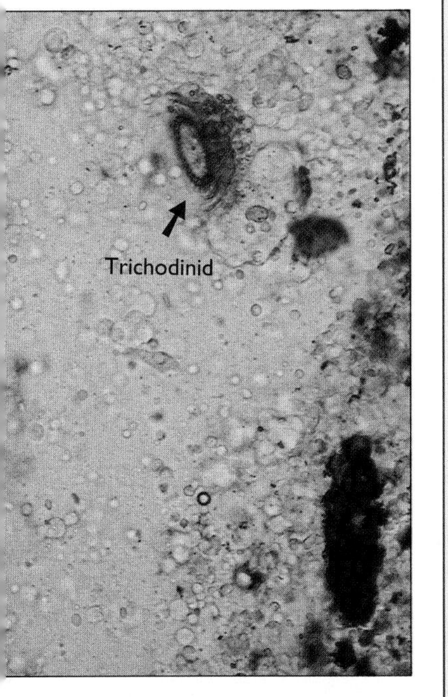

Above: A skin scraping reveals several parasites, including the long shape of the skin fluke, *Gyrodactylus* and the distinctive ciliated trichodinid protozoans.

WHAT IS A MONOGENETIC FLUKE?

The monogenetic flukes (or Monogenea) are flatworms that are usually found as external parasites of aquatic animals. They often have an obvious hooked attachment organ at the hind end and they have simple, direct life cycles involving sexual reproduction and only one type of host. Some monogeneans are egglayers; some are livebearers.

Monogeneans are usually a few millimetres in length, and hence just visible to the naked eye, although more clearly so with a x10 magnification hand lens. As external parasites, they often feed upon skin and gill tissue, but usually only cause real harm if present in large numbers. Formalin or organophosphorus insecticides (both of which must be used with extreme care) are usually successful in treating monogenean infestations of fish.

The monogenetic flukes are often grouped with the digenetic flukes (Digenea – see page 181) and referred to collectively as the Trematoda (trematodes), of which there are about 2400 species in all.

Below: A close view of the livebearing fluke *Gyrodactylus* shows the hooks of an unborn fluke within its body. Adults are 0.5-1mm(0.02- 0.04in) long.

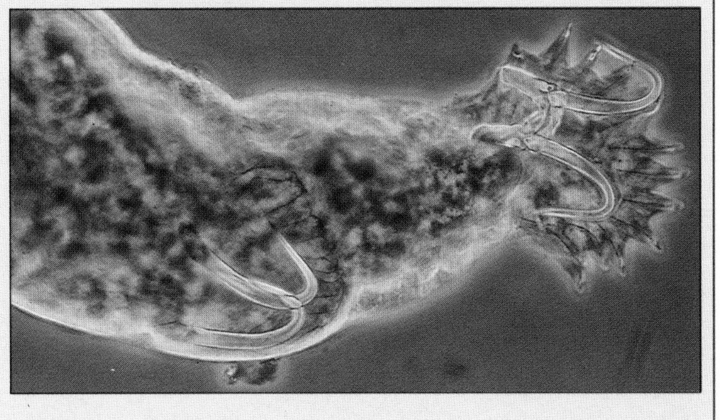

SWIMBLADDER DISORDERS

Caused by

Various factors, including sudden changes in temperature, although other factors, such as microbial infection, may be involved in some instances.

Obvious symptoms

Although an affected fish usually appears in reasonable condition, it experiences difficulty in maintaining its position in the water. It may show a 'list' to one side, for example, or even float on its side or back at the water surface or remain sunk on the bottom of the tank or pond.

If the fish exhibits sporadic floating problems, such that it swims normally for some of the time, this may indicate air-gulping and not a swimbladder problem.

Occurrence of the disease

Swimbladder disorders may occur in a variety of situations, often spontaneously in previously healthy stock. Other fish in the same pond or aquarium may remain unaffected. Certain 'fancy' varieties of goldfish, such as moors, veiltails and orandas, are particularly prone to this condition, and often have misshapen swimbladders.

Treatment and control

Since the exact causes of this problem are poorly understood, recommending a reliable treatment is difficult. Air-gulping occurs when fish feed greedily at the surface and suck in air with the food.

Right: A Ryukin Veiltail fancy goldfish swimming at an unnatural angle, clearly a sign of a swimbladder disorder. Ironically, some of the most beautiful fancy goldfishes are prone to these problems because their extreme shape affects the arrangement of the swimbladder within the body.

Below, right: This fancy goldfish is suffering from displacement of one of the swimbladder chambers by an abnormally enlarged kidney, causing the fish to consistently float on its side.

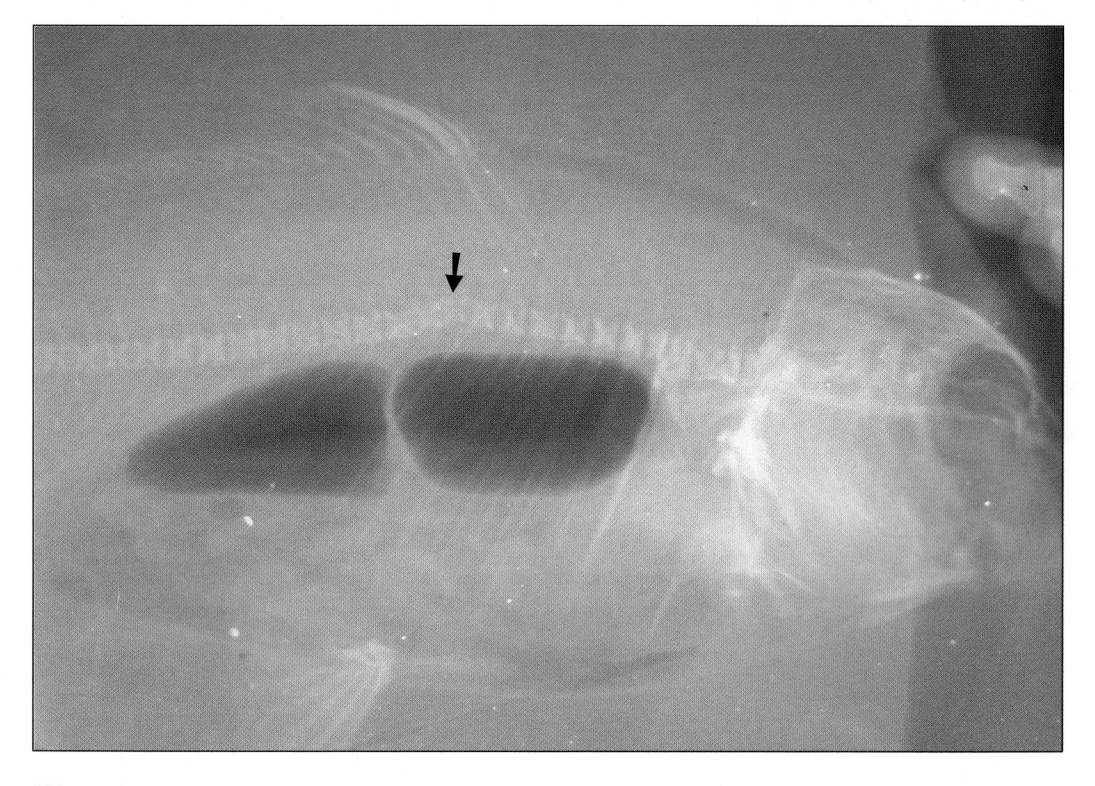

Below: This X-ray shows the typical two-chambered arrangement of the swimbladder in cyprinid fishes, such as goldfishes and koi. The X-ray was originally taken to confirm a suspected spinal fracture, clearly evident (arrow).

One solution is to briefly hold the food (e.g. flakes or pellets) just beneath the water surface so they quickly sink when released and have to be eaten in mid-water.

If the buoyancy problem is not due to air-gulping, then it could be a more serious condition affecting the swimbladder. Try moving the fish to shallow water, say 13cm (5in) depth. In the case of goldfish and other coldwater fish, slowly increasing the water temperature by about 5°C (9°F) sometimes brings about an improvement. Dosing the tank with an antibacterial remedy and/or aquarium salt (up to 1gm salt per litre, assuming the fish are salt-tolerant) may be effective in some cases. A veterinarian may be willing to X-ray the affected fish in order to visualize any damage or derangement of the swimbladder chamber(s). If swimbladder over-inflation is discovered, it may be possible for the vet to aspirate the excess gas, although this does not always bring about a permanent cure. Despite attempts at treatment, many cases of swimbladder problem fail to improve. If the affected fish seems very distressed or is unwilling or unable to feed then it might be kinder to have it put down.

TUMOURS AND LYMPHOCYSTIS

Caused by

Tumours are caused by a variety of environmental factors, such as chemical pollution, and perhaps by certain viral infections. Some tumours may be inherited from the parent fish. Lymphocystis is caused by a viral infection. (See also *Fish pox*, page 116.)

Obvious symptoms

Tumours are unusual growths or swellings and may occur in any part of the body. Those on the skin and fins are usually quite obvious, although similar growths may occur among the internal organs, sometimes causing firm, noticeable swellings to the general body shape.

Lymphocystis causes raspberry- or cauliflower-like growths on the skin and fins. It often begins as small, white 'cysts', which gradually increase in size over a period of weeks or months.

Goldfish are susceptible to a condition known as kidney enlargement disease which, as its name suggests, causes the kidney to become greatly swollen. Affected fish develop an enlarged ventral region and scale erection, giving rise to dropsy-like symptoms (see page 104). There may also be loss of balance.

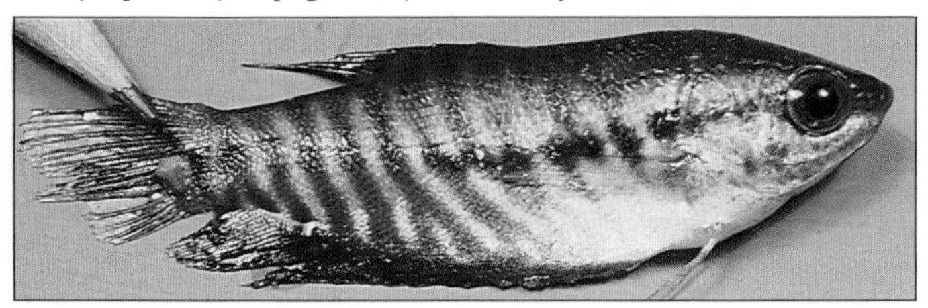

Above: These small rough lumps on the skin and fins are typical of lymphocystis. They are quite distinct from the smooth waxy growths caused by fish pox.

Above: A goldfish with a tumour on its flank. It is possible for a veterinarian to surgically remove such external growths, but there is a chance they may recur.

Below: A fish coping with a tumour on its 'chin'. Many such growths do not necessarily cause any discomfort or are infectious, but it is wise to isolate such fish.

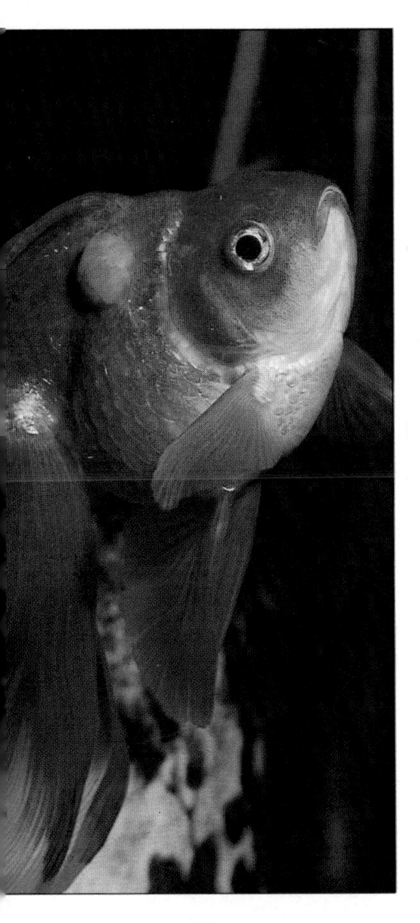

This life-threatening condition is caused by myxozoan parasites (*Hoferellus* species) that infect the kidney tubules.

Above right: This tumour in the area of the vent may cause the fish to suffer; it is best to painlessly destroy affected fish if treatment proves ineffective.

Occurrence of the disease

The tumours of fish are rarely very infectious, and, in many instances, the factors responsible for their occurrence are poorly understood.

Fish suffering from lymphocystis may transmit the disease to other fish via abrasions of the skin. Infection can remain dormant and undetected for some time. Although most often a disease of marine and brackish-water fish, it may also occur in certain groups of freshwater fish, notably cichlids and gouramis. Conversely, it does not affect cyprinids (carps, barbs and relatives) or catfish. It is rarely fatal, being unsightly rather than highly pathogenic.

Kidney enlargement disease has been linked with overcrowded conditions and poor hygiene.

Treatment and control

It is a good idea to segregate fish showing growths of an uncertain nature into a hospital tank and to monitor their condition. There is no effective treatment for tumours, lymphocystis or kidney enlargement disease. Some veterinarians may be willing to attempt surgical removal of external growths from badly affected fish. However, even after surgery, such growths may recur.

Badly affected stock should be painlessly destroyed and fish showing signs of 'growths' should not be purchased.

POLLUTION, TUMOURS AND DISEASE

Tumours, or growths, on or in the body of fish, can be caused by a range of factors. However, there is some evidence to suggest that environmental pollution may bring about a higher incidence of tumours in populations of wild fish. For example, in a study in North America the incidence of fish with tumours in a 'polluted' river was around five percent, compared to a one percent incidence of tumours in fish in a 'clean' river. The pollutants in question included heavy metals, arsenic, oil and insecticides. It was suggested that these pollutants may have been involved in triggering off tumour production in the fish.

Pollution may also have similar effects on outbreaks of infectious diseases. It has been shown experimentally that trout exposed to low levels of copper are more susceptible to certain bacterial infections, for example, and that more general pollution may increase susceptibility of eels (*Anguilla anguilla*) to bacterial infection, and of salmon and related fish to some viral and bacterial diseases. Thus, environmental pollution can have varied and long-lasting effects, in addition to the obvious, short-term consequences often reported in the media.

ULCER DISEASE AND HAEMORRHAGIC SEPTICAEMIA

Caused by
Various bacteria, including *Aeromonas*, *Pseudomonas* and *Vibrio*.

Obvious symptoms
Lesions, ulcers or sores on the body, reddening at the base of the fins and the vent, loss of appetite and darkening of coloration are all symptoms of infection. Dissecting an affected fish often reveals an accumulation of fluid in the body cavity and haemorrhages among the internal organs. Ulcers may become secondarily infected with fungus. In very acute disease outbreaks, however, fish may die showing very few obvious external symptoms.

Occurrence of the disease
The bacteria responsible for this type of disease are often very common in the aquatic environment or as low-level 'latent' infections in otherwise healthy fish. However, outbreaks usually occur only in fish which are in poor condition for some reason, such as recent importation, rough handling or overcrowding. In a poorly maintained pond or aquarium, the disease can have devastating effects, as diseased fish release pathogenic bacteria into the water, which in turn infect other fish.

Below left: Ulcerated lesions on a carp. The broken skin is vulnerable to further infection.

Below: Prevent small lesions, such as those on the snouts of these fancy goldfish, from developing further by adding an antibacterial remedy to the tank.

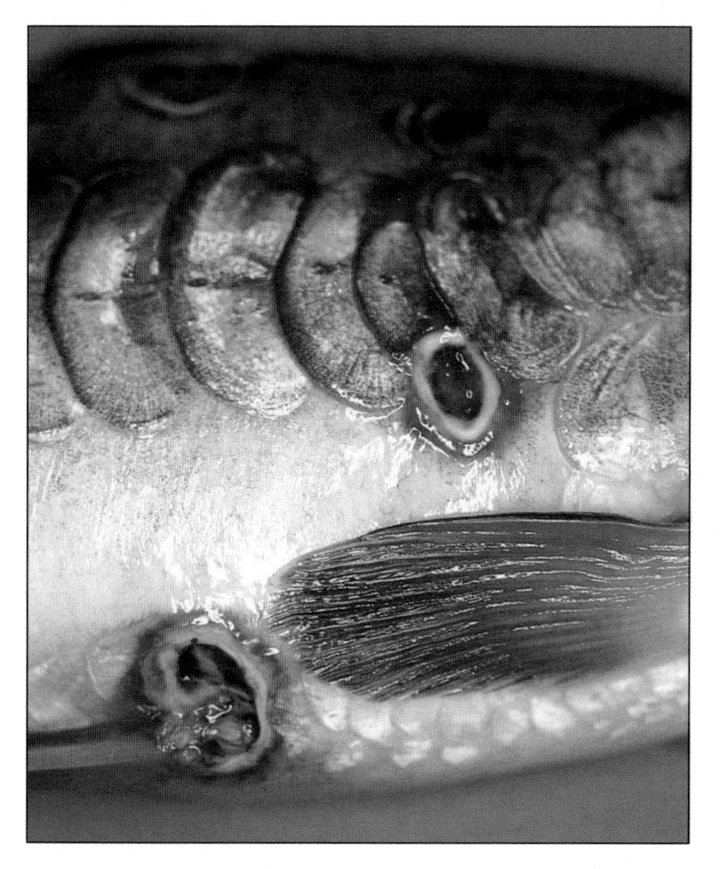

Above: When an ulcer reaches this condition, it is vital to isolate the fish and treat it promptly with nifurpirinol or a suitable antibiotic. Seek expert advice.

Right: Ulcer disease is commonly seen on goldfish, but other fish, such as tropical freshwater and tropical marine fish, can also suffer from similar problems.

Below right: Signs of bacterial haemorrhagic septicaemia in the internal organs, with pronounced reddening and an accumulation of bloodstained fluid.

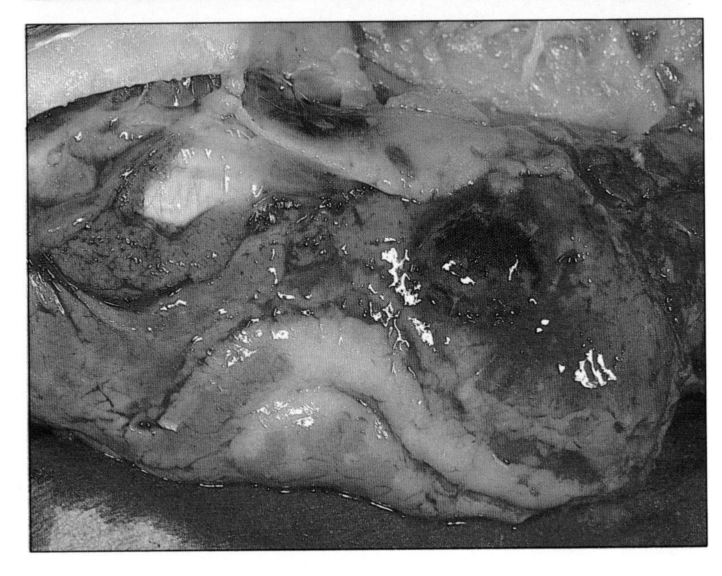

Treatment and control

If they are still feeding, offer fishes showing early signs of this disease a full course of antibiotic medicated flake or pelleted food. Be sure to give the other, apparently healthy, fishes the same medicated foods.

Where the disease is more advanced, and the fish are reluctant to feed, place the obviously affected fish in an isolation tank and treat them by adding nifurpirinol or, if unavailable, antibiotics or a similar antibacterial to the water (see Chapter 7).

Large fish can be injected with a suitable antibiotic solution, and a topical antiseptic applied to any obvious lesions at the same time. Maintain such fish in isolation while symptoms persist and consider adding salt to the water to ease their osmotic stress.

The best method of long-term control is to identify and eliminate the factors responsible for bringing on the disease.

Left: Lesions such as this harbour many millions of pathogenic bacteria, which pass into the water and can infect other fishes in the same aquarium or pond. When treating for early signs of infection, ensure that apparently healthy fish benefit from the treatment.

Below: Physical damage and abrasions, here shown on a koi, can lead to secondary infections with fungus and bacteria. To contain the infection apply a topical antiseptic and/or use an antibacterial in the water (see Chapter 7 for details).

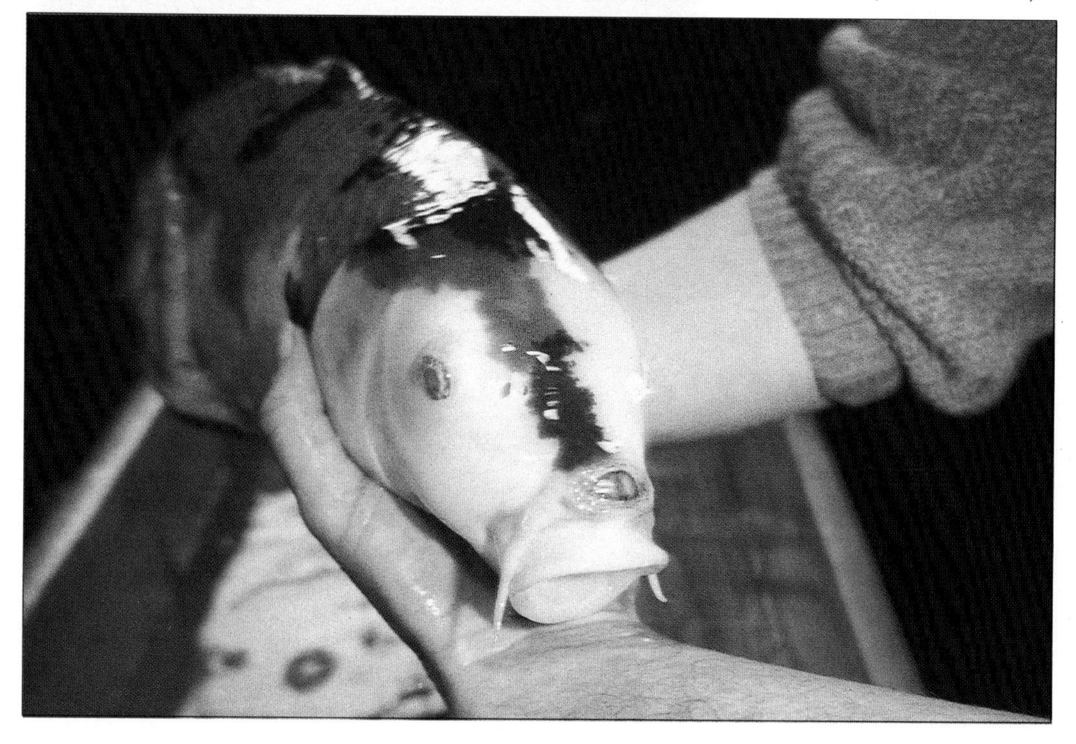

WHAT IS A BACTERIUM?

A bacterium is a microscopic, single-celled organism, often measuring between 0.5 to 10 microns (thousandths of a millimetre). As a group they occur everywhere there is life. Bacteria can be shaped like a sphere ('coccus'), a rod ('bacillus') or a spiral ('spirillum'), and can be arranged singly, in chains or in clusters. They are usually visible at a magnification of x400 to x1000 on an ordinary light microscope, although groups or colonies of bacteria can be visible to the naked eye and show recognizable features.

Bacteria have a tough, rigid outer cell wall. This is not impermeable, since bacteria must absorb all their food in solution through the cell wall. Bacteria never contain the green photosynthetic pigment chlorophyll, although many contain other such pigments. Some bacteria have a slimy outer capsule, and there may be whiplike flagella to enable them to move through liquids. On the other hand, many simply drift in air or water currents.

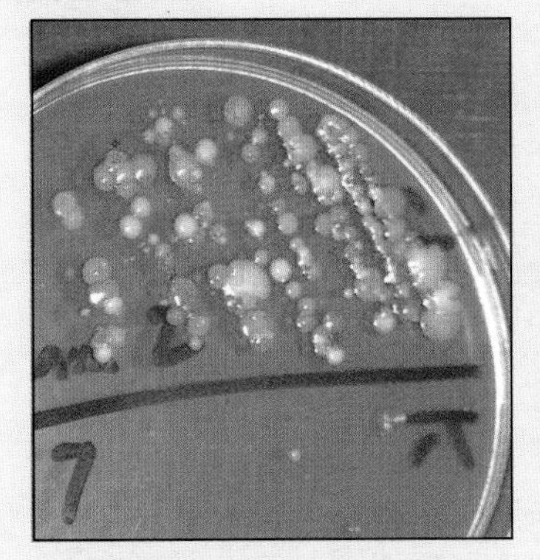

Above: To study bacteria, they are first cultured on nutrient medium in flat dishes. These blobs are bacterial colonies.

Bacteria usually reproduce by splitting into two – so-called binary fission – and in some this may occur once every 15-30 minutes. This means that, given favourable conditions, one such bacterium could form well over 150,000,000,000,000 bacteria in 24 hours. Usually, however, this potential is not realized, either because the food supply runs out or toxic waste products build up.

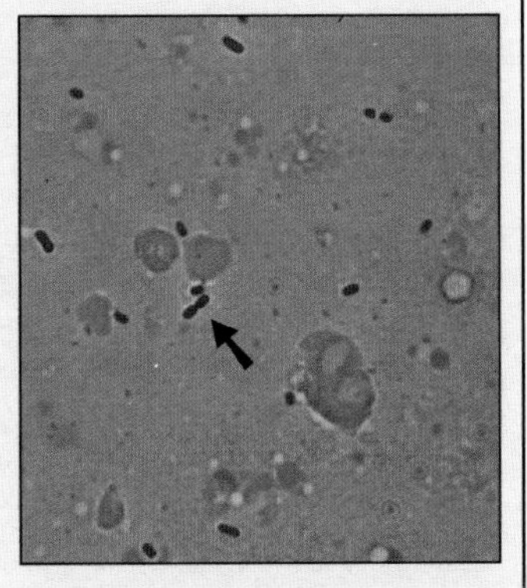

Above: The small dark dots are single and paired *Aeromonas* (ulcer disease) bacteria, each measuring about 1-2 microns.

Bacteria are very numerous, and generally very tough. Just a pinch of garden soil contains many millions of bacteria, for example, and some can survive freezing, intense heat, drying and even some disinfectants. Bacteria usually survive such adverse conditions by forming resistant spores, which can remain active for a number of years.

Many bacteria are helpful to man, and are responsible (at least in part) for the decay of dead plants and animals, an unceasing process that releases nutrients to be recycled again. Bacteria also help treat our sewage, produce cheese, yoghurt and beer, and certain resident intestinal bacteria (the so-called 'gut flora') help many animals digest their food. The nitrification process, by which ammonia and nitrites are converted to less harmful nitrates, is the result of bacterial activity.

Some bacteria, however, may cause diseases, and in man such diseases include typhoid, leprosy and plague. Many bacterial infections can be successfully treated using antibiotics and similar drugs, and even vaccination. Many of the pathogenic (i.e. disease-producing) bacteria responsible for fish diseases are quite common in the aquatic environment, but only cause outbreaks of disease when fish are kept in unsuitable conditions.

The form of some common bacteria (not to scale)

Cocci (spheres) singly, in clusters and in chains

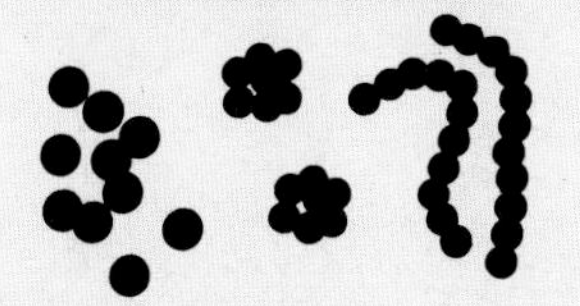

Spirillum (In a spiral)

Bacilli (Rods)

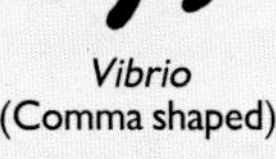

Vibrio (Comma shaped)

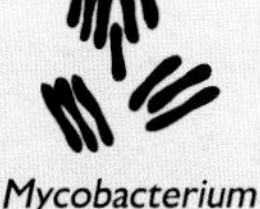

Mycobacterium (Very small rods)

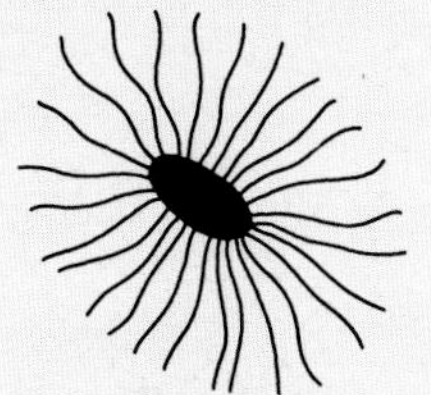

Peritrichous (Flagella all round)

Flexibacter (Long, thin rods)

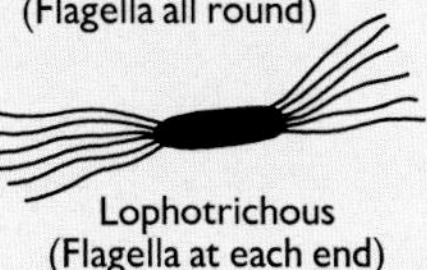

Lophotrichous (Flagella at each end)

VELVET DISEASE AND CORAL FISH DISEASE

Caused by

Parasitic single-celled organisms known as dinoflagellates. *Amyloodinium* on marine fish; its freshwater equivalent is *Piscinoodinium* (commonly referred to as Oodinium).

Obvious symptoms

Yellow-grey coating to the skin and fins. Fish may scrape against rocks and show increased gill movements. In advanced cases, the fish go off their food and lie motionless in the water; skin may peel away in strips. Although this disease may be confused with other conditions, such as white spot, velvet-infected fish often look as if they have been sprinkled with gold dust – hence the alternative common name of 'gold dust disease'.

Occurrence of the disease

The parasite moves from fish to fish in the form of flagellated (tailed) spores, which can live away from the fish for at least 24 hours and probably much longer – perhaps for several days. It is most often introduced with new fish and may then develop into a serious problem. Heavy infestations on the gills may kill fish without causing any obvious signs of the disease. The disease appears to be particularly persistent among killifish, some anabantoids, goldfish and marine coral fish. Long-term control may be complicated if the parasite establishes itself within the fish's gut, where it may escape treatment.

Right: Velvet disease usually appears as a fine dusting of spots like a sprinkling of gold dust, clearly shown here on this Indian glassfish (*Chanda ranga*).

Below: A closer look at velvet disease on this rasbora shows how the clustering of the dots may cause confusion with white spot disease. Each dot is a parasite.

The life cycle of *Piscinoodinium* and *Amyloodinium*, the parasites responsible for velvet and coral fish disease

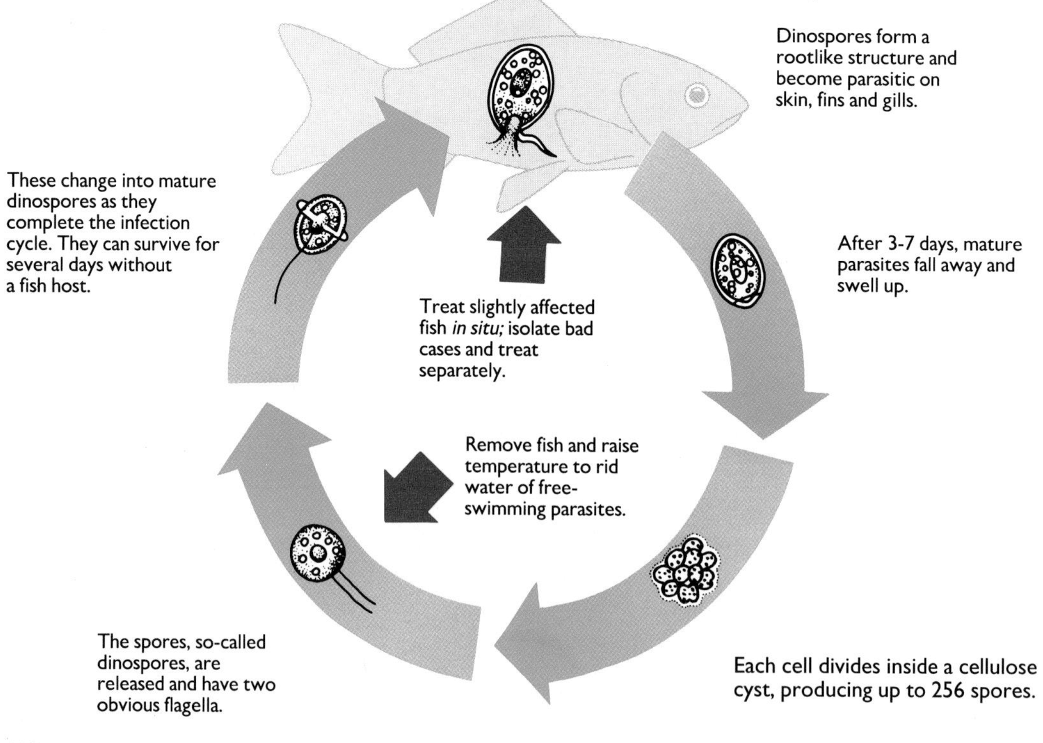

Treatment and control

In the case of freshwater velvet (*Piscinoodinium*), a proprietary brand of velvet remedy, white spot remedy, or a broad spectrum anti-parasite treatment should effect a cure. If the affected fish are salt-tolerant, then a prolonged salt treatment (1 teaspoon salt per 5 gallons) can be used to eradicate the parasites. During serious outbreaks it may help to keep the aquarium relatively dark. Darkness will prevent the parasites from photosynthesising and also delays dinospore development.

With marine fish, a three- to four-week course of copper treatment is recommended (see Chapter 7). Removing all the fish to a treatment tank, and then raising the infected tank temperature to 30-32°C (86-90°F) for three weeks usually eliminates most of the free-swimming velvet parasites. High temperature in itself does not kill the parasites, but it does speed up their life cycle, shortening the length of time they can survive without their fish hosts.

The speed at which the infestation develops, and the severity of the problem which may ensue, highlights the importance of quarantine for all new fish, especially tropical marines.

Left: A light dusting of *Piscinoodinium* on the skin and fins of a rosy barb (*Barbus conchonius*).

Below: This heavily infested white cloud mountain minnow (*Tanichthys albonubes*) is beyond the stage at which treatment is effective and will be a source of infection for the whole aquarium. Such fish should be isolated from the main aquarium and ideally humanely destroyed.

WHAT ARE DINOFLAGELLATES?

Piscinoodinium, the velvet parasite familiar to freshwater hobbyists, and *Amyloodinium*, the causative organism of coral fish disease in marine aquariums, are both dinoflagellate organisms.

In general, dinoflagellates are most common as free-living members of the plankton in the ocean, although some forms are parasitic. Dinoflagellates typically have hairlike flagella for locomotion, and some contain the green pigment chlorophyll. Since they share features with both algae and flagellate protozoans, they can be classified with either, although often with algae. In common with algae and other plants, some dinoflagellates can manufacture their own food by photosynthesis, although some also engulf tiny food particles.

Amyloodinium and *Piscinoodinium* typically reach up to 0.2mm (0.01in) in size, and are thus just visible to the naked eye.

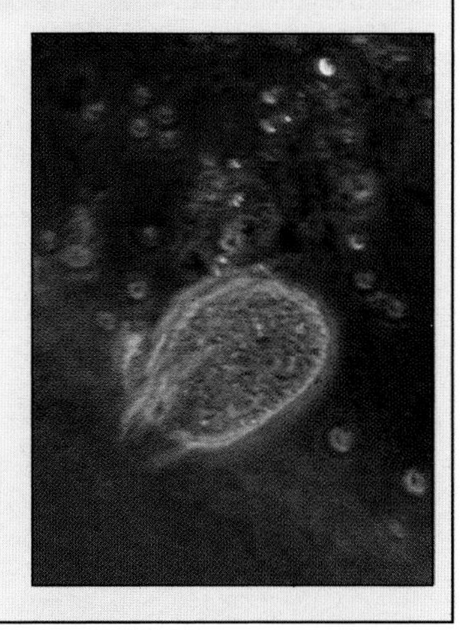

Right: A single *Piscinoodinium* parasite attached to the gill filament of its fish host by a 'root' structure. Actual size usually less than 0.15mm (under 0.01in)

WASTING DISEASE AND FISH TUBERCULOSIS

Caused by
Infection with acid-fast bacteria, such as *Mycobacterium* or *Nocardia*.

Obvious symptoms
Fish suffering from this disease often have an emaciated, hollow-bellied appearance, with coincident loss of appetite and loss of colour. They may exhibit other symptoms, including 'pop-eye', finrot, body ulcers and listless behaviour. Many small nodules, or tubercles, (about the size of a pinhead) usually occur within the internal organs of affected fish, although a fungal parasite called *Ichthyophonus* may also produce similar nodules.

Right: Chronic infections with *Mycobacterium* can eat into the internal organs. Here, healthy liver tissue has been replaced with tuberculous nodules.

Below left: Emaciation and pale shallow lesions in the skin may suggest infection with *Mycobacterium*. Isolate fish as soon as they show such symptoms.

Below: Tuberculosis nodules among the body organs of a goldfish. Culturing the causative organisms using standard bacteriological methods can be difficult.

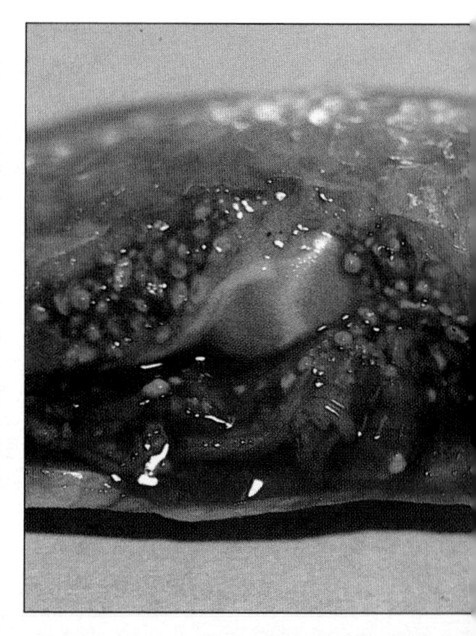

Right: Cannibalism can transfer many diseases between fish, including fish tuberculosis. To prevent infection, it is vital to remove all dead or dying fish.

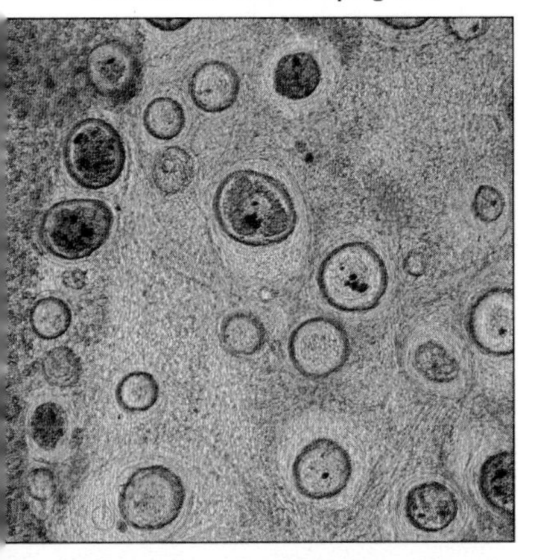

Left: This Mexican livebearer exhibits symptoms of wasting disease: listless behaviour, emaciated body, and cloudy eyes. Special bacteriological staining techniques may be used to confirm whether a mycobacterial infection is the cause.

Occurrence of the disease

The disease is probably passed from fish to fish by feeding on infected material, although the passage of *Mycobacterium* from parent fish to offspring is also possible, especially in livebearers. Feeding fish with other infected fish, or materials from infected fish, is therefore an important source of infection. As with many diseases, apparently healthy fish may harbour the infection without ever showing signs of the disease. However, an outbreak may be brought on if fish carrying the infection are subjected to poor environmental conditions.

Treatment and control

Buying apparently healthy stock and caring for them correctly will help prevent outbreaks of this disease. Naturally, it is important not to feed fish on live or dead fish from infected sources.

Segregate any fish suspected of being infected with *Mycobacterium* or *Nocardia* and, if their condition does not improve, painlessly destroy them. Drugs are available which may be of some use in treating this disease, and if expensive or particularly unusual fish are involved, contact a local veterinarian for advice. Drugs which have been used with some success include: sulphafurazole, doxycycline and minocycline (see Chapter 7). Your veterinarian will administer these drugs by intramuscular injection, or may prescribe suitable alternatives.

The fungal parasite *Ichthyophonus* is very difficult to treat, but it is a disease of uncertain significance to ornamental fishkeepers.

The mycobacteria that cause wasting disease in fish are capable of infecting humans (typically causing a persistent skin rash on the fingers or hands). Therefore, it is essential to avoid contact with uncovered (and especially broken) skin when handling infected stock or contaminated equipment or water. Following an outbreak, thoroughly disinfect equipment and other facilities (see page 198).

WATER QUALITY PROBLEMS

Caused by

Incorrect or fluctuating water conditions (particularly pH, hardness, specific gravity and temperature), general poor hygiene, excessive levels of nitrite or ammonia, presence of other toxins in the water and/or low levels of dissolved oxygen.

Obvious symptoms

The effects of unsuitable water conditions may vary from small changes in 'normal' behaviour to large-scale fish losses. In acute cases, the majority of the fish in a pond or aquarium all exhibit unusual behaviour suddenly. Symptoms often include peculiar swimming behaviour, rapid gill movements, periods of inactivity interspersed with darting movements, gasping at the water surface, pop-eye and cloudy eyes. Death may follow. At a more chronic level, less pronounced water quality problems may just make the fish appear 'off-colour', unwilling to feed and more susceptible to diseases such as finrot and fungus. Under such conditions, fish may also appear unwilling to breed and egg/fry survival may be poor.

Occurrence of the problem

Water quality problems are a root cause of many fish diseases. In some cases adverse water conditions have a direct effect on the fish's health, such as ammonia poisoning causing gill and skin damage. In other situations poor water quality exerts an indirect effect by impairing the fish's immune system resulting in an increased susceptibility to infection.

Water quality problems are especially common in newly set up or poorly maintained aquariums. Fish bowls and very small aquariums are particularly prone to water pollution due to their small water volume and typical lack of biological filtration.

Treatment and control

Water quality problems can be avoided by sensible, regular aquarium or pond maintenance along with regular monitoring of water conditions. Naturally, it is important to avoid overstocking

Above: Blood congested in fancy goldfish fins can indicate water quality problems, such as too low a temperature and high ammonia.

Left: A discus flipping over in poisoned water and showing signs of a white spot-like infection and rapid deterioration of the skin, scales and fins. Acute water quality problems often have such sudden drastic effects on fish in ponds and aquariums.

Right: Chronic poor water conditions can debilitate fish, here a fancy goldfish, and make them fall foul of a whole range of different infectious diseases.

and overfeeding and to take care to prevent the entry of paint fumes, toxic sprays, garden run-off, tobacco fumes and other toxins into the water. Always condition all new tapwater with a reliable water treatment before use in the aquarium. Before buying fish check whether they are compatible with the water conditions within your aquarium or pond.

Naturally, proper pond and aquarium care will prevent this type of problem occurring. However, if fishes begin to show signs of suffering in unsuitable water conditions, the situation may be alleviated by carrying out an immediate water change of about 50-75 percent, topping up the pond or aquarium with new, conditioned water at the correct temperature. In some situations, particularly acute cases, it may be better to remove the fish to a separate tank full of conditioned water at the appropriate temperature and then investigate the underlying causes.

DISEASES IN THE WILD

Outbreaks of disease can be a serious problem, affecting fish in ponds, aquariums and fish farms. Under such conditions of close confinement, where stocking densities are high and water quality conditions are perhaps less than ideal, fish can fall foul of a range of potential pathogens. However, as we have discussed throughout this book, disease *prevention* is in the hands of the hobbyist and fish farmer.

What is not often realized is that outbreaks of diseases also occur in wild fish populations, too. Sometimes these are brought on by natural overcrowding, poor water conditions, or factors that we do not yet fully understand. Every spring in the UK, for example, populations of fish such as roach (*Rutilus rutilus*), bream (*Abramis brama*) and similar 'coarse' fish in small ponds and lakes suffer from outbreaks of external parasites, such as fish lice (*Argulus*), white spot, sliminess of the skin, etc. Sometimes only a few dozen fish die in a given locality; sometimes many thousands perish.

It need not happen in the same pond or lake each year, and why it happens at all is not completely understood. It is likely that fluctuating or rising water temperatures, poor water quality, overcrowding, poor physiological condition at the end of winter, and the onset of spawning may all make the fish more susceptible to these infestations. In addition, rising temperatures favour a rapid build-up of parasite numbers.

For some years, a disease called 'perch ulcer disease' has been affecting perch (*Perca fluviatilis*) populations in the UK, and it has been estimated that one million perch died of this disease in Lake Windermere, in the Lake District of England, during the summer of 1976. It is now thought that perch ulcer disease is caused by *Aeromonas salmonicida*, the bacterium implicated in furunculosis of trout and ulcer disease of goldfishes and koi.

Therefore, the problems caused by outbreaks of diseases are not restricted to ponds, aquariums and fish farms, as such problems also occur in the wild. However, since outbreaks of infectious diseases in ponds and aquariums are often brought on by poor environmental conditions, perhaps equivalent outbreaks of disease in natural waters should be taken as important environmental warnings for us all to heed.

WHITE SPOT DISEASE OR 'ICH'

Caused by
Ichthyophthirius multifiliis in freshwater tanks and *Cryptocaryon irritans* in marine systems. Both are ciliate protozoans, which may be accompanied by other skin and gill parasites.

Obvious symptoms
Small, white spots on the skin, fins and gills. Each spot is an individual parasite that lies just beneath the fish's transparent skin epithelium. The spots measure up to 1mm (0.04in) across and in heavy infections several spots may aggregate to form irregular white patches. Heavily infected fish may look as if they have been sprinkled with salt or sugar grains, and they may scratch against rocks and gravel, and show increased gill movements. In ponds, heavily infected fish often collect in shallow water, among the plants. Secondary bacterial infections are common.

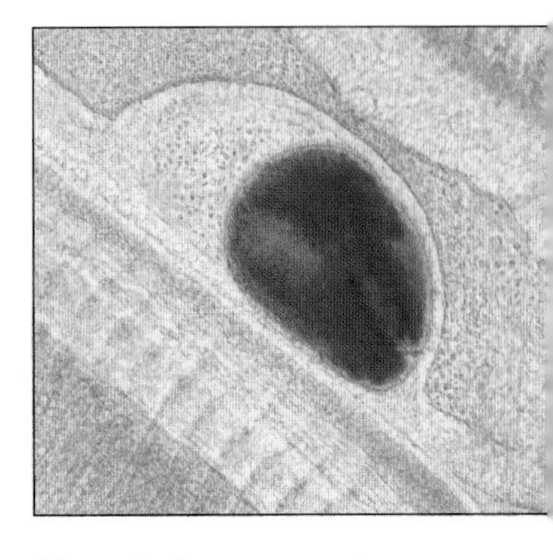

Above: A white spot parasite attached to a gill filament. At low levels, such infections can pass unnoticed yet act as a disease reservoir for other fish.

Occurrence of the disease
Ichthyophthirius is capable of infecting virtually all freshwater fish including coldwater and tropical species. In contrast, the marine fish parasite, *Cryptocaryon*, is restricted to warm water temperatures (above 18 °C, 64 °F).

Although not closely related to one another, these two parasites share remarkably similar life cycles (see page 169) that involve parasitic and free-living stages. Mature parasites, which have been feeding on host tissues, break through the skin epithelium and fall away from the fish. On the floor of the aquarium or pond they settle to form a round cyst which is the reproductive stage. Within

Below: The clear signs of white spot on an impressive cichlid, *Geophagus hondae*. White spot can infect just about every species of pond and aquarium fish.

Above: The larger, more rounded spots of white spot disease serve to distinguish it from the smaller spots of velvet and coral fish disease (page 158).

each cyst the parasite divides many times, eventually producing many hundreds of infective stages, or 'swarmers'. These exit the cyst and swim off in search of a fish host. If they do not find a host within a few days, they die.

The time taken for the life cycle to turn full circle (i.e. from fish to fish) varies with temperature, being much faster in tropical than in temperate conditions. In *Ichthyophthirius*, the complete cycle takes about three or four days at 21°C(70°F); at 10°C(50°F), this period is extended to at least five weeks. At low temperatures, the parasite may lie dormant for some considerable period. *Cryptocaryon* is more dependent on high temperatures, and rarely causes problems below 20°C(68°F).

The relatively high stocking levels in aquariums and garden pools greatly increase fish to fish transmission of these parasites.

Below: A fish clearly suffering from white spot and fungus. It is important to treat white spot promptly to prevent these secondary infections arising.

Treatment and control

While attached to the host, the white spot parasite is situated beneath the outermost layer of the skin, and hence is immune to the treatment. Consequently, chemical control, i.e. by adding chemical treatments to the water of an infected aquarium or pond, is usually aimed at the free-living stages. A number of excellent proprietary white spot treatments are available for use in fresh water. Some freshwater fish, such as clown loaches and scaleless catfishes, are a little sensitive to some of the above treatments, however. Allowing a pond or aquarium to be left fish-free for at least seven days at 20°C (68°F) or above also usually eliminates the white spot parasite from the system.

In seawater systems, use one of the proprietary treatments that have been developed for marine fishes, noting that some such treatments are usually copper-based and toxic to marine invertebrates. Leaving the marine aquarium fish-free for 8 weeks will usually eliminate the parasite from the system (marine invertebrates do not suffer from white spot and so may be left in the aquarium during this period). This isolation period is far longer than that required for freshwater white spot, due to *Cryptocaryon*'s long-lived cyst stage.

Prompt treatment is vital for effective control of white spot. Since this disease may be introduced with new fish, a suitable period of quarantine and a reliable treatment are recommended. White spot can also be introduced with plants and live food, if they have been in contact with other fish during the previous few days. In the future, it may be possible to vaccinate fish against white spot disease (see page 201).

Left: A red emperor snapper (*Lutjanus sebae*) with the whitish nodules caused by *Cryptocaryon irritans* covering its fins and head. The parasites are largely immune to chemical treatment while they are attached to the fish.

The life cycle of *Ichthyophthirius multifiliis* and *Cryptocaryon irritans*, the cause of white spot disease

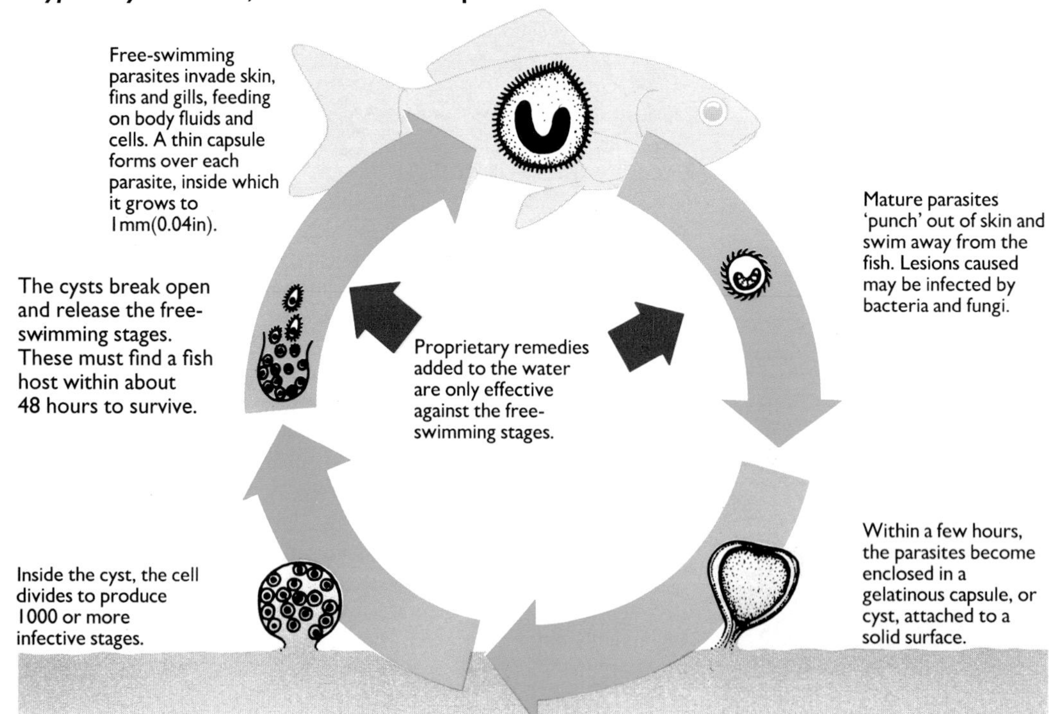

GUPPY DISEASE

As the name suggests, guppy disease is a condition most often associated with guppies (*Poecilia reticulata*). It is caused by *Tetrahymena*, a tiny ciliate protozoan. Infestations of other skin parasites may occur together with *Tetrahymena*.

The visible symptoms are very similar to those of white spot disease; small white 'spots' occur on the skin and in the muscle, each up to 1mm(0.04in) in diameter. These spots are usually accompanied by a falling away of the outer layer of skin, protruding scales and unusual swimming behaviour. In some instances, the parasites may eat deep into the muscle and even enter the bloodstream.

Tetrahymena most often causes a problem in newly imported fish or those in poor condition for some other reason, such as poor or inadequate nutrition, excessive organic debris, or incorrect water quality. Although guppies are quite hardy fish, they, too, can become vulnerable to disease if they are kept in unsuitable conditions. They thrive in warm, moderately hard, quite alkaline water and become more susceptible in very soft, acid water which is relatively cool.

Adding a proprietary brand of white spot treatment to the affected tank can be effective, although well-established infections may be more difficult to control, and require several treatments to be successful.

Below: The frayed tail and 'ragged' appearance caused by a combination of guppy disease, slime disease and finrot.

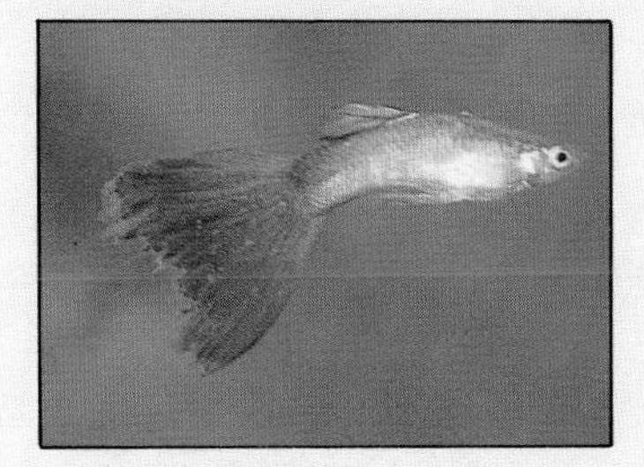

WHAT IS A PROTOZOAN?

A protozoan is a tiny animal that consists of a single complete cell. Most are microscopic, measuring perhaps 10 to 100 microns (thousandths of a millimetre), and are usually visible at a magnification of x100 to x200 on an ordinary light microscope. Some protozoans are large enough to be visible to the naked eye, such as the white spot parasite (*Ichthyophthirius*) when attached to the fish. Protozoans are especially common in damp or watery environments, although some can survive drying by forming resistant spores or cysts. Most protozoans are free living, although some of the 30,000 or so known species are parasitic. Such parasites may have a simple life cycle involving one host, or a more complex life cycle involving intermediate hosts and vectors (see the panel on page 127).

Protozoans typically possess one nucleus within their cell (but sometimes more), a contractile vacuole for regulating water content, and perhaps cilia or flagella for feeding and movement through liquids. They often reproduce by splitting into two – so-called binary fission – although some forms may also involve a sexual process. Some protozoans may contain a photosynthetic pigment, such as chlorophyll, and may be difficult to distinguish from plants. Most require dissolved nutrients or perhaps solid food particles, such as bacteria, to 'ingest'.

There are four very distinct groups of protozoans:

● **Flagellates** (Mastigophora)
These possess one or more flagella for propulsion. *Euglena* is a well-known free-living example of this group, and *Trypanosoma* is a blood parasite which causes sleeping sickness in man and equivalent symptoms in fish and some other animals.

● **Ciliates** (Ciliophora)
These have large numbers of short cilia over the body surface, and use these for movement and food collection. *Paramecium* and other infusorians are free-living ciliates. *Ichthyophthirius* is a ciliate parasite which causes white spot in fish.

● **Amoebas** (Sarcodina)
These simple protozoans flow across surfaces using jelly-like extensions to their bodies. Most amoebas are free living and are found in freshwater ponds, although a few may cause disease problems in man and animals.

● **Sporozoans** (Sporozoa)
These are parasites of other invertebrates and vertebrates, and may have quite complex life cycles. Micro- and myxosporidian parasites of fish, such as those that cause nodular diseases, are included in this group.

Various drugs and chemicals are available for treating protozoan infestations, although they can be more difficult to treat internally than when they occur on the exterior, especially in fish.

Below: Special staining reveals the curved nucleus in this single *Ichthyophthirius* cell.

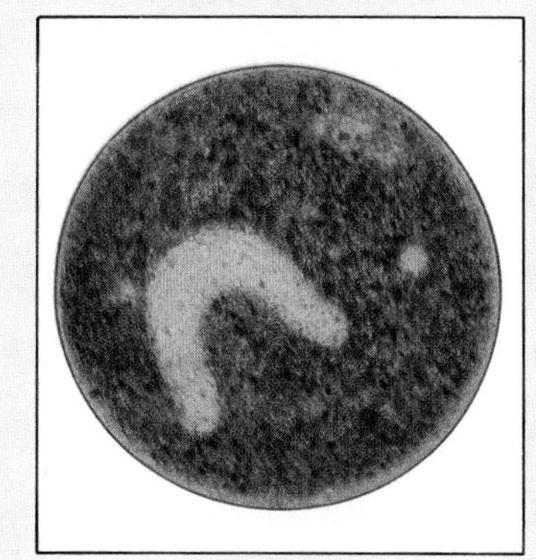

WORMS IN THE BODY CAVITY

Caused by
Various helminth ('worm') parasites, such as cestodes (tapeworms) and nematodes (roundworms). (See also *Worms in the intestine*, page 172 and *Yellow grub*, page 178).

Obvious symptoms
Low-level infestations probably pass unnoticed and do little harm. Heavy infestations may cause a swollen belly, impaired swimming behaviour, damage to internal organs and rupture of the body wall. Tapeworms are elongated, white, ribbon-like parasites up to several centimetres long. Signs of external segmentation may be present. Nematodes are similar in appearance to small pieces of cotton or string, often yellow-brown in colour. Both these parasites may live free in the body cavity or encapsulated in white or off-white cysts.

Occurrence of the disease
These parasites are most common in newly imported or wild-caught fish. Because of their complex life cycles – involving two or three hosts – these parasites rarely build up to serious levels in the pond or aquarium.

Treatment and control
Fortunately these parasites are rarely a problem to fishkeepers, since reliable treatment is difficult. Do not buy fish showing gross symptoms of worm infestation and avoid using live foods such as cyclops and 'water fleas' (the intermediate host of many of these parasites) unless they originate from a fish-free water source.

Above: Heavy infestations of worms in the body cavity cause swelling and may make fish more susceptible to predation as they struggle to swim properly.

Below: In extreme cases, the internal organs of the fish host can be displaced, as shown here by a very heavy infestation with the tapeworm *Ligula intestinalis*.

The typical life cycle of tapeworms, such as *Ligula* and *Schistocephalus*, that infest the body cavity of fish

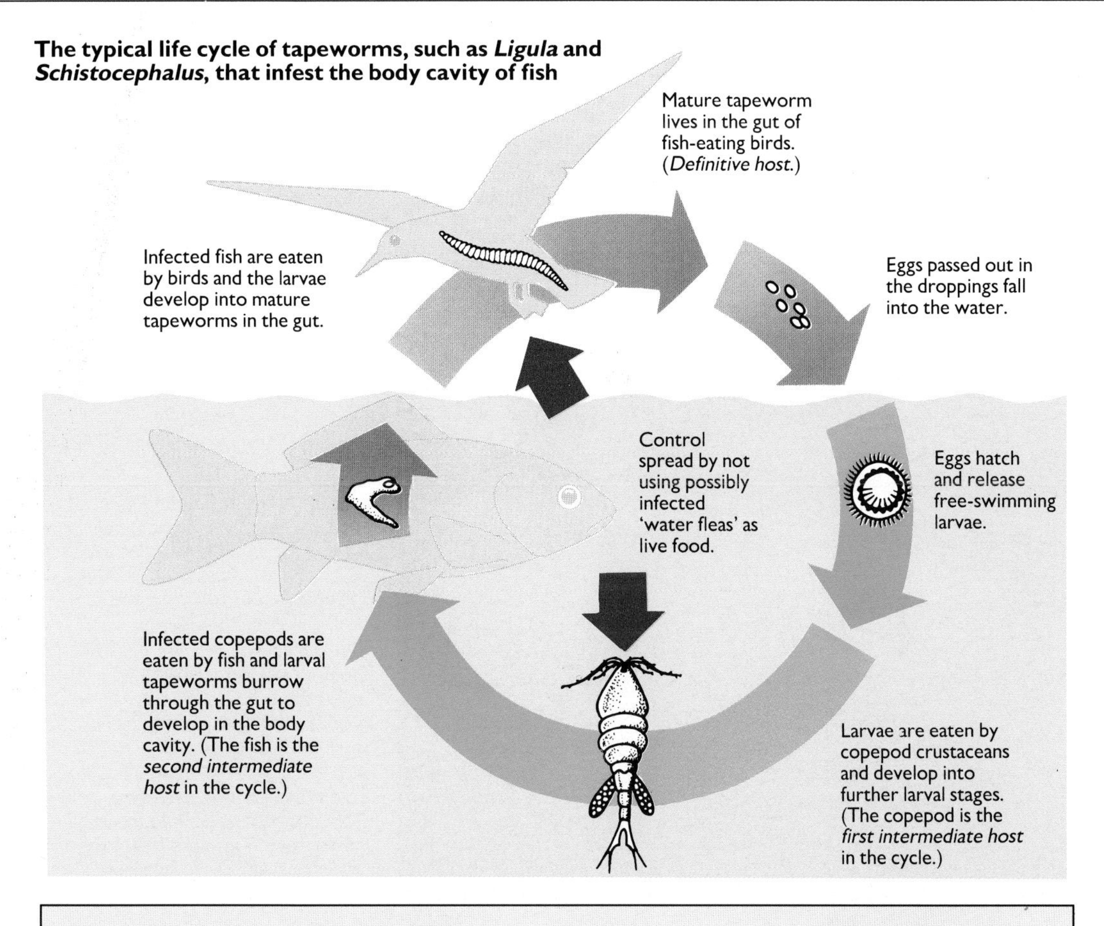

WHAT IS A CESTODE?

There are approximately 1500 species of cestodes (Cestoda), or tapeworms, and all are parasites. Adult tapeworms generally occur as parasites in the digestive system of vertebrate animals, including fish. They have complex indirect life cycles, involving one or more intermediate hosts. These may include invertebrates, such as copepods and shrimps, and vertebrates, such as fish. Thus, fish may harbour both adult tapeworms in their intestine and/ or larval tapeworms, usually among the visceral organs in the body cavity.

Adult tapeworms may reach over 10m(33ft) in length, although in fish the adults and larvae usually measure from a few centimetres to, perhaps, 40cm(16in) in length. Cestodes have no mouth or gut, but absorb nutrients directly through the outer covering, or tegument. The adult parasites are usually made up of a chain of segments, or proglottids. At the front end of the chain is an attachment organ consisting of suckers, hooks or muscular depressions. Each proglottid contains a complete set of reproductive organs, and large numbers of parasite eggs are released into the outside world inside mature proglottids that pass out with the faeces of the definitive host.

Anthelmintic drugs are commonly used to remove the familiar adult tapeworms from dogs and cats. Such drugs are available for use on fish, but are only effective in eliminating gut-dwelling tapeworms and not those residing within the fish's body cavity.

Below: The larval stage of the tapeworm *Schistocephalus solidus* from the body cavity of sticklebacks. 2-4cm(0.8-1.6in).

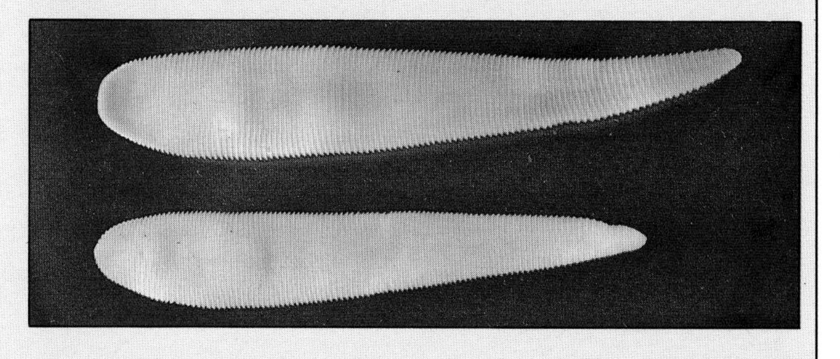

WORMS IN THE INTESTINE

Caused by
Various helminth ('worm') parasites, such as cestodes (tapeworms), nematodes (roundworms) and acanthocephalans (spiny-headed worms). Digenetic flukes also occur here, but are rarely pathogenic in fish.

Obvious symptoms
Except in the case of very heavy infestations, where the fish may appear thin or grossly distended, with the parasites perhaps

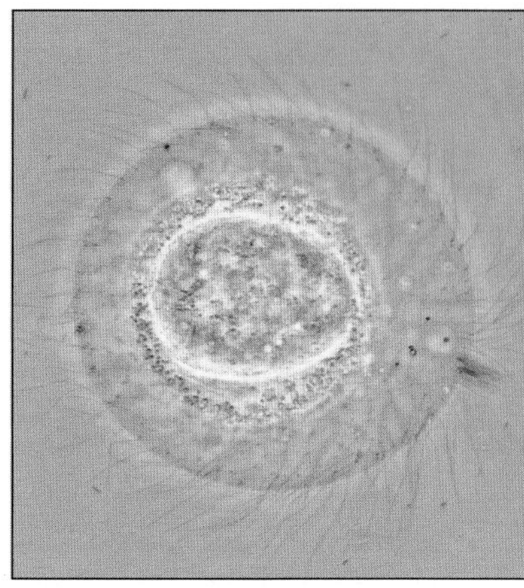

Above: A newly hatched larva of the tapeworm *Bothriocephalus acheilognathi*. This must be eaten by an intermediate host, such as a copepod, to develop further.

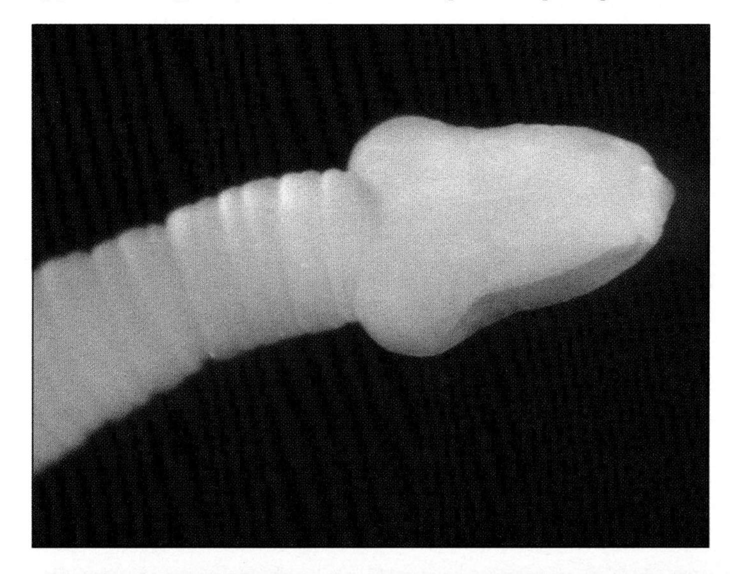

Left: The anterior region of the tapeworm *Bothriocephalus acheilognathi*, showing its modification for attachment to the host intestinal wall.

Below: An adult tapeworm, *Bothriocephalus acheilognathi*, from a carp. This tapeworm has a fish/copepod/fish life history and causes problems on fish farms.

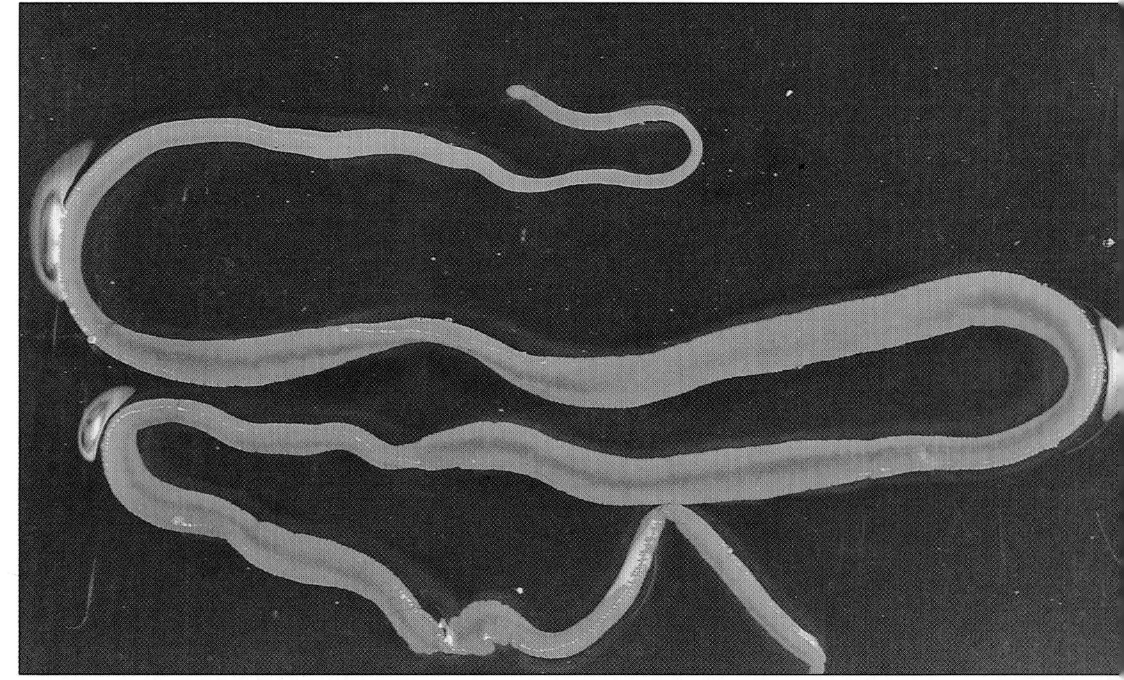

protruding from the vent, obvious symptoms of infestation are usually lacking. These parasites are most common in wild-caught and newly imported fish. Tapeworms usually appear as flat, white, often tangled pieces of ribbon, sometimes divided into segments. Nematodes have a characteristic wormlike appearance, like a short piece of cotton or string. They may be red-brown in colour and reach several centimetres in length, although some forms are very small. Acanthocephalans are usually yellowish white and measure 1-2cm(0.4-0.8in) in length. They are flattened or cylindrical, with a retractable spiny proboscis at the 'head' end, which attaches the parasite firmly to its host's intestine.

Below: Adult digenetic flukes, such as this *Bunodera luciopercae*, also occur in the intestines of fish but are rarely pathogenic. Note the two suckers for attachment. The actual length of this fluke is usually around 2-5mm(0.08-0.2in).

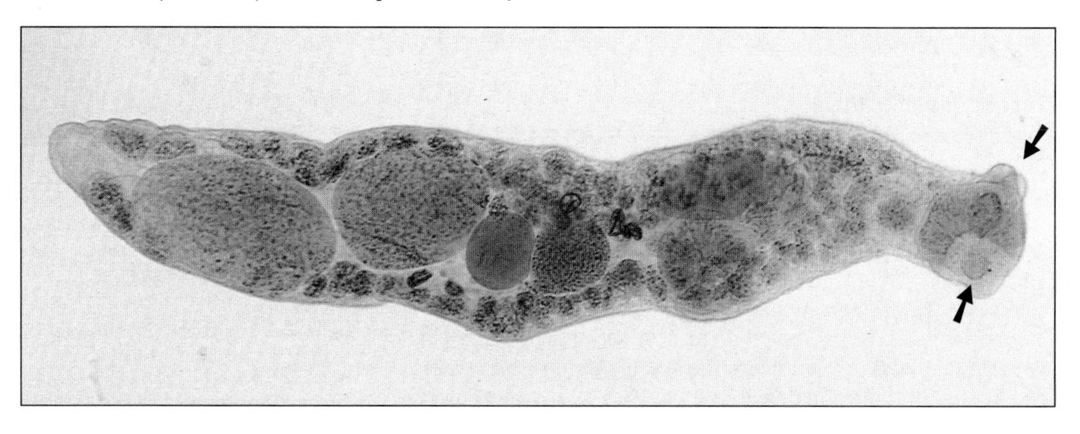

The life cycle of *Bunodera*, a digenetic fluke that infests the intestine of fish

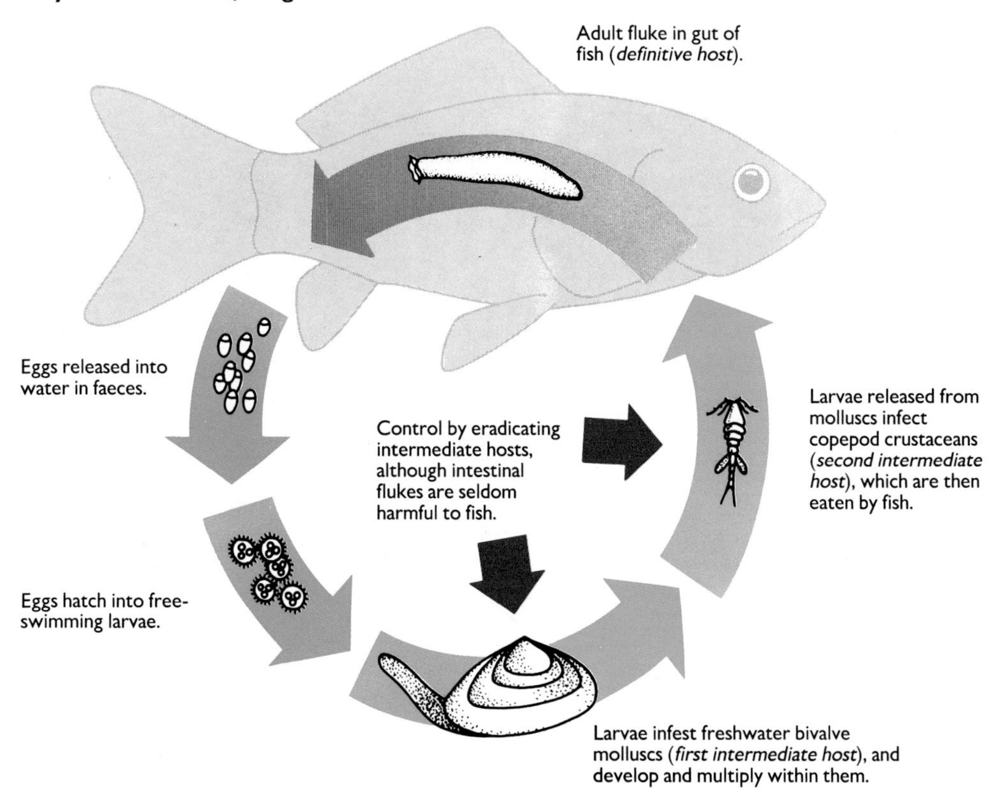

Occurrence of the disease

In most instances, fish hobbyists will not be aware of this type of infestation in their fish. These parasites usually have very complicated life cycles, involving two or three hosts. Hence, they do not easily build up to dangerous levels in a pond or aquarium.

One exception to this is the nematode *Camallanus*. This is a common parasite of many tropical livebearing fish, and can cause problems when heavy infestations are present in small fish. Although *Camallanus* nematodes usually require a copepod ('water flea') intermediate host to convey the juvenile worm from fish to fish, this parasite is also able to pass from fish to fish directly – at least for several generations. This can result in heavy infestations in some situations.

Treatment and control

Since most infestations remain undetected, and probably do little harm, treatment is not usually necessary. In the simplified environment of a pond or aquarium, the other hosts necessary for the completion of the parasite life cycle are usually absent. 'Worm' parasites present in the intestine of newly imported or wild fish will eventually die and reinfection will be unlikely.

However, *Camallanus* nematodes may multiply within the confines of an aquarium – even in the absence of their intermediate host. In such situations, anthelmintic treatment may be desirable. (The treatment of intestinal helminth parasites is discussed more fully in Chapter 7.)

Above: A heavy infestation of the acanthocephalan *Pomphorhynchus laevis* in a fish intestine. Each such spiny-headed worm can reach several centimetres long.

Below: A freshwater *Gammarus* shrimp acting as an intermediate host by harbouring the orange larvae of the acanthocephalan *Pomphorhynchus laevis*.

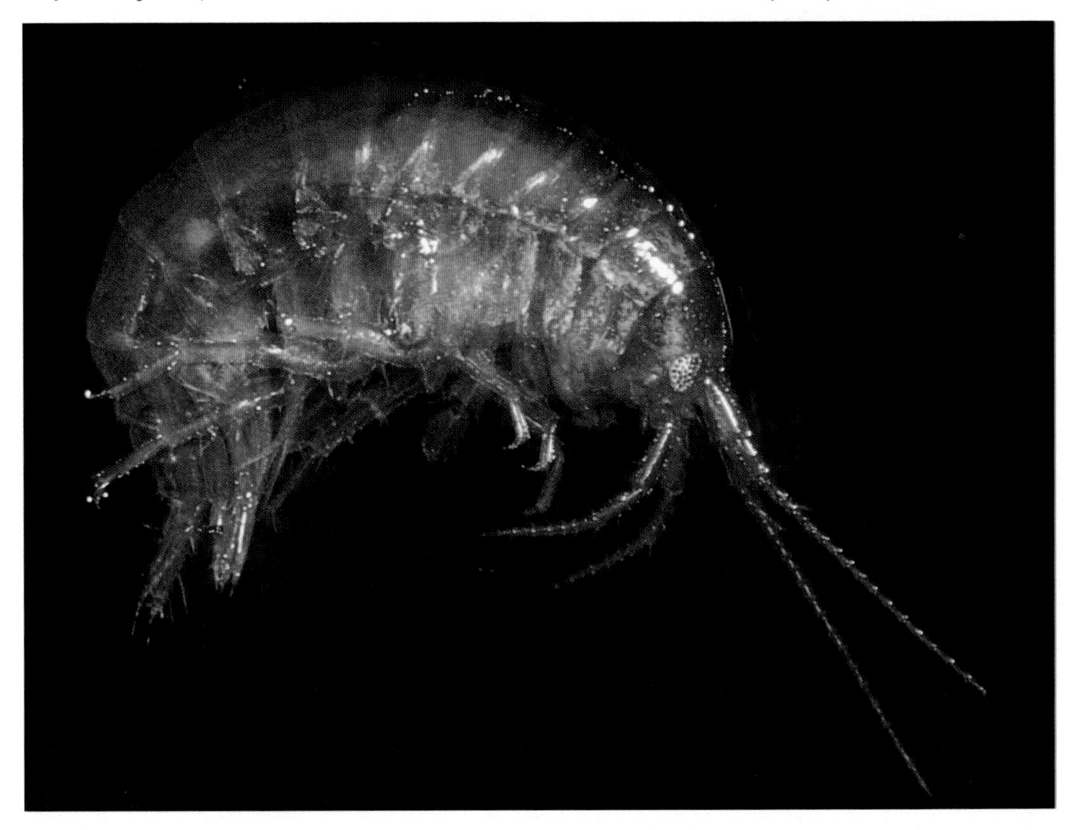

The life cycle of a spiny-headed worm, such as *Pomphorhychus laevis*

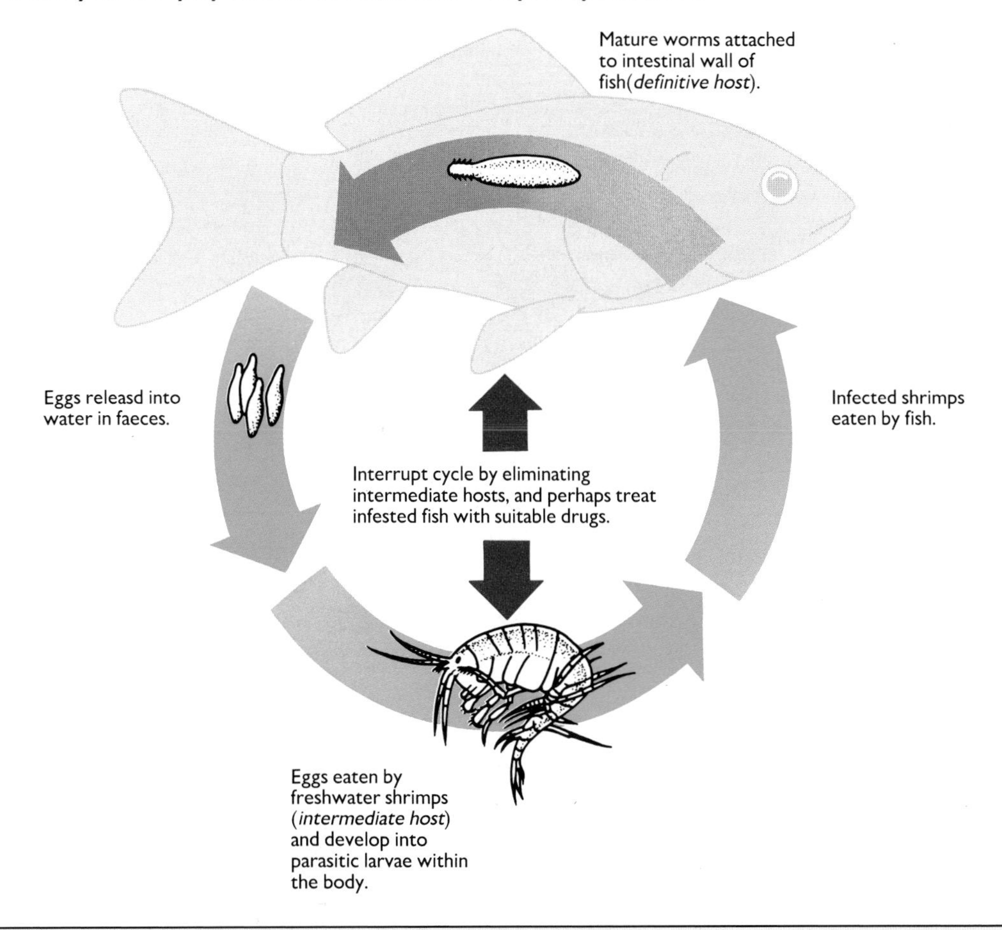

WHAT IS A SPINY-HEADED WORM?

The spiny-headed worms, or acanthocephalans (Acanthocephala), are a group of approximately 400 species of parasites. They live as adults in the digestive system of vertebrates, including fish, using the long spiny, but retractable, proboscis to attach themselves to the gut wall of their host. Acanthocephalans have no gut or mouth, but absorb nutrients directly through their body wall. They are usually about 2-3cm(0.8-1.2in) long, although some species can grow much longer than this.

The male and female parasites mature and mate, the female then releasing large numbers of 'eggs', which reach the outside world with the host's faeces. There is always at least one intermediate host in the life cycle, commonly an invertebrate such as a shrimp, which usually becomes infected by feeding on mature eggs.

Because of their complex, indirect life cycles, spiny-headed worms rarely cause problems to captive fish, although, as with many such parasites, wild-caught fish may harbour them when first obtained. After some months, however, these parasites are likely to die and disappear.

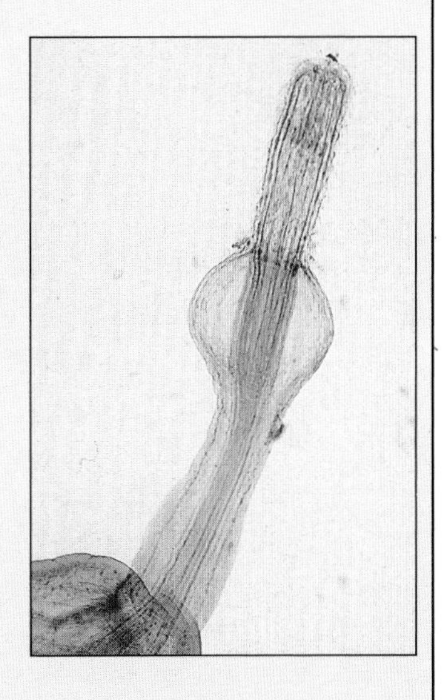

Right: The long spiny proboscis of the fish acanthocephalan *Pomphorhynchus laevis*, which can penetrate the intestinal wall and bore into vital organs.

The possible infection cycles of *Camallanus*, a roundworm that infests the intestine of fish

Right: The red-brown worms protruding from the vent of this upturned fish are *Camallanus* nematodes, each up to 2cm(0.8in).

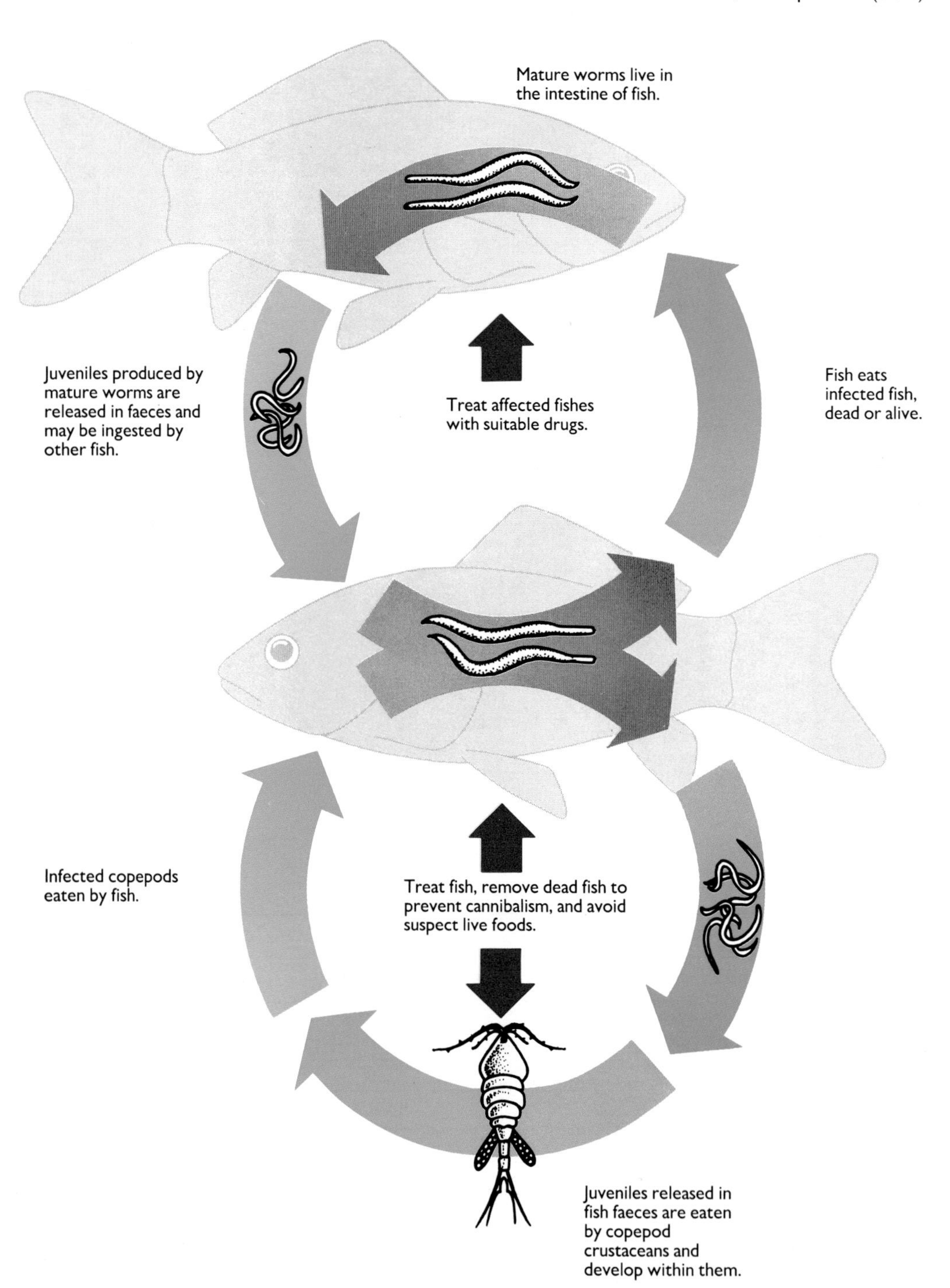

WHAT IS A NEMATODE?

Nematodes, or roundworms, are a large group of more than 10,000 species of free-living and parasitic threadlike animals. Many are just a few millimetres to perhaps one or two centimetres in length, although some can reach 1m(39in).

Nematodes usually have separate sexes, and the life cycles of parasitic species can range from very simple to extremely complex. Some nematodes are also important pests of crops, farm animals, dogs, cats and even man. Treatment is usually possible using an appropriate anthelmintic drug and, if applicable, by breaking the life cycle by eliminating any of the intermediate hosts and/or improving general hygiene to remove the infective stages.

Roundworms have a well-developed gut, and some parasitic forms have strong mouthparts and a muscular 'throat'. When present in large numbers, which is relatively unusual under captive conditions, they can be quite debilitating to their host by feeding on tissues and blood. In fish, as with other animal hosts, nematodes may be found in a number of locations, including the digestive system, body cavity and swimbladder.

Below: The 'head' end of a *Camallanus* worm, with tiny 'jaws' that cause considerable damage to the host's intestines.

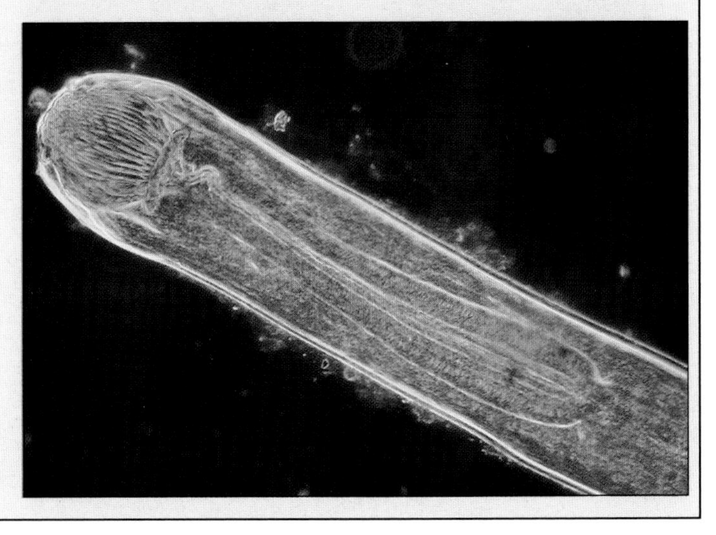

YELLOW GRUB, BLACK SPOT AND EYE FLUKE

Caused by
Larval stages of digenetic fluke parasites, such as *Clinostomum*, *Posthodiplostomum* and *Diplostomum*.

Obvious symptoms
In yellow grub and black spot, appropriately coloured cysts occur on the body and fins. These cysts often measure up to 2mm(0.08in) across and contain the tiny larval parasite. Similar cysts may also occur among the internal organs. (Here, there is a possible confusion with wasting disease, page 162). Small numbers of these cysts do little harm, although large numbers are unsightly and may be dangerous to small fish.

The parasite responsible for eye fluke lodges in the lens, humour (liquid) or retina of the eye and may do considerable damage if present in large numbers. The extent of this damage ranges from lens cloudiness and rupture to blindness. However, low-level infestations will probably occur unnoticed.

Below: Very small black spots on the skin of marine fish, as shown on this surgeonfish, are more likely to be caused by an infestation with a small turbellarian flatworm, which is a close relative of the skin flukes. Additional symptoms normally include fin flicking and scraping against rocks. This problem can usually be treated by using an organophosphorus insecticide, such as metriphonate (see Chapter 7).

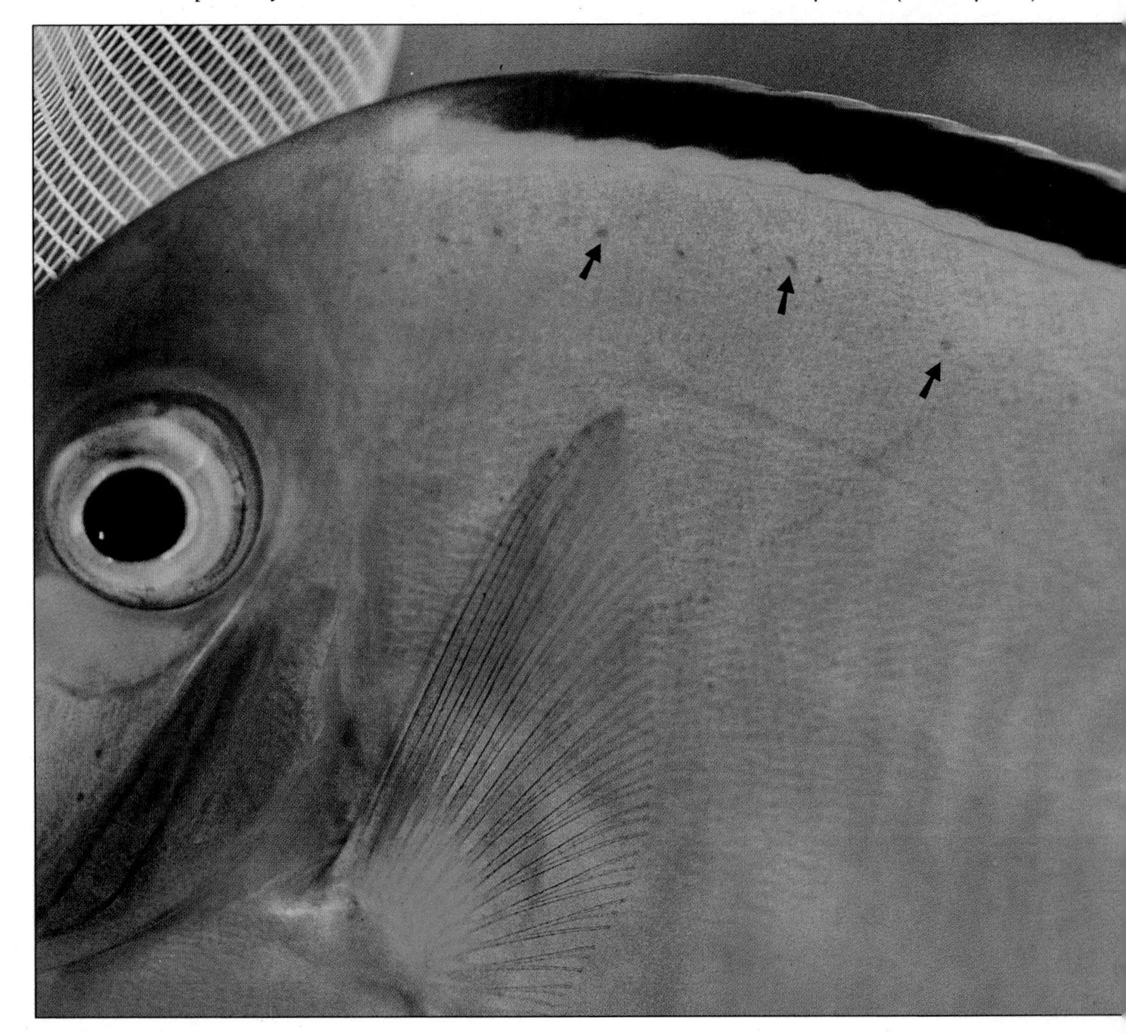

Below: An eye fluke, *Diplostomum spathaceum*, removed from the lens of a fish. The actual size of this larval parasitic stage is about 0.5-1mm(up to 0.04in) long.

Below: Eye flukes can cause cloudiness and even lens rupture but, because of their complex life cycle, such parasites are rarely a problem in ponds or aquariums.

Above: The typical signs of black spot, here on a freshwater fish, the dace (*Leuciscus leuciscus*). Such low-level infestations do little harm and usually disappear with time.

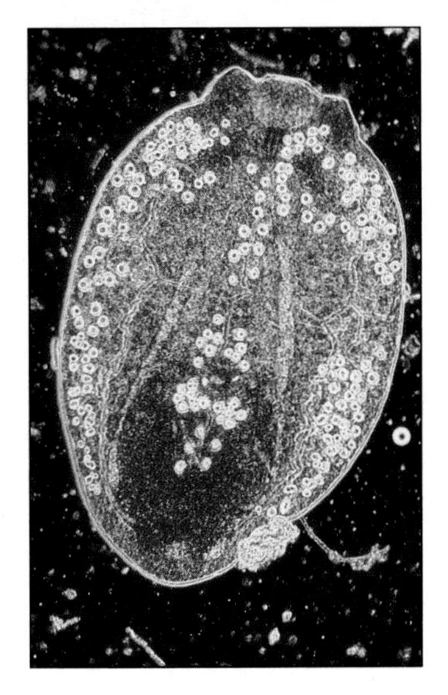

Occurrence of the disease

This disease most often occurs in newly imported fish and occasionally in pond fish. The parasites mature in the intestines of fish-eating birds or mammals. Here, they produce eggs which, on entry into water, usually infect aquatic snails. Fish most often become infected by tiny invasive larvae leaving the snail and penetrating a suitable fish host.

In most pond and aquarium situations, it will be impossible for such a complex life cycle to be completed. Therefore, these parasites rarely build up and cause problems to fish hobbyists. Although the larvae in the fish are long-lived, they will eventually die, to be consumed by their host's immune responses.

Right: A close up view of a black spot on the skin of a freshwater fish. Each shiny cyst contains a larval parasite. The black coloration is produced by the host as a reaction to infection.

Treatment and control

Treatment is rarely necessary and is, in any case, difficult, if not impossible. Avoid buying fish with obvious heavy infestations of these parasites, and discourage fish-eating birds from visiting garden ponds to prevent the life cycle being completed.

The generalized life cycle of the digenetic flukes that cause yellow grub, black spot and eye fluke

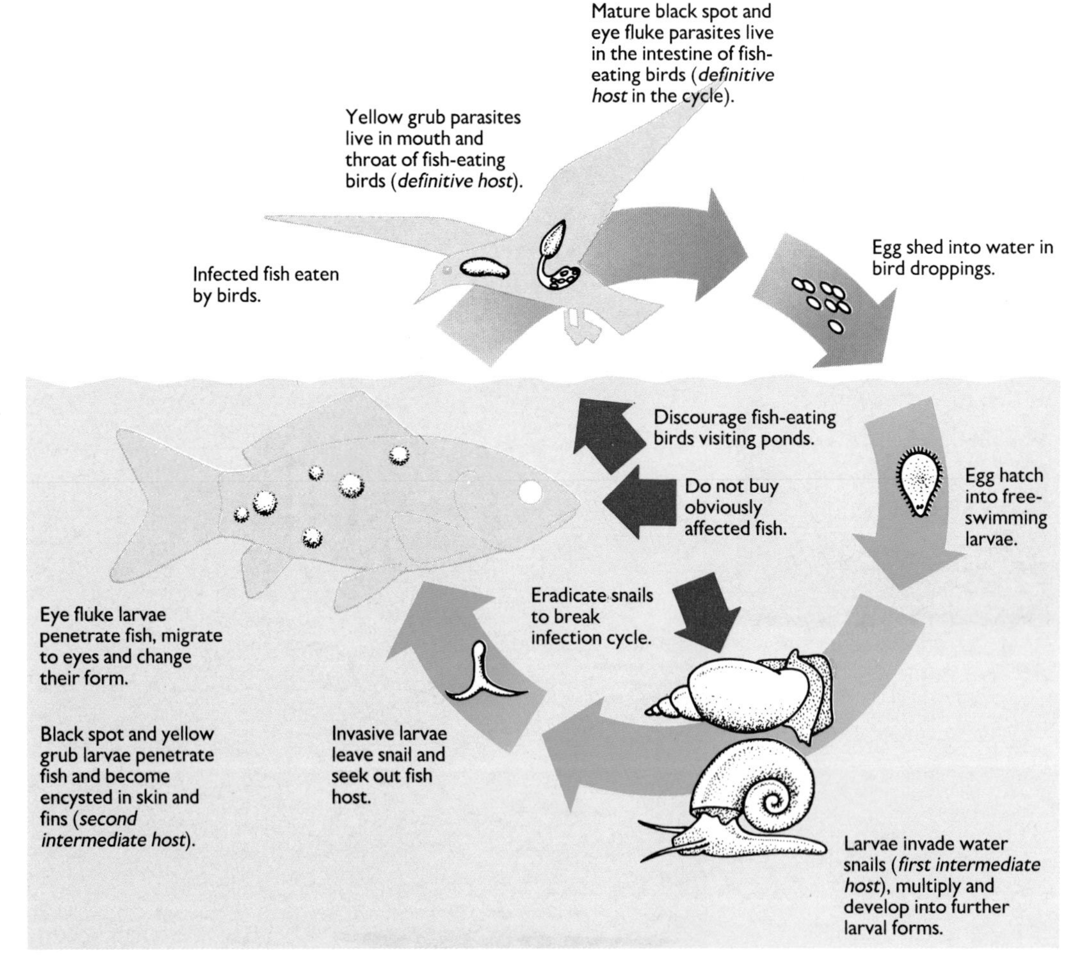

WHAT ARE DIGENETIC FLUKES?

The digenetic flukes (or Digenea) are parasitic flatworms with complex indirect life cycles involving vertebrates, including fish, and invertebrate hosts, such as molluscs and copepod crustaceans. The adult parasites are hermaphrodite (i.e. contain both male and female sex organs) and mature in the intestines of the definitive host (see the panel on page 127). Sexual reproduction produces large numbers of eggs, which pass to the outside world in the faeces. One or more intermediate hosts may then be involved in the life cycle, before a definitive host becomes infected. Each stage in the life cycle must undergo a period of development (or even multiplication) in the intermediate host or hosts, otherwise the life cycle cannot continue.

Fish may harbour adult digeneans in the gut, blood system or bladder, or larval digeneans (often referred to as metacercariae) in the eyes, fins, skin, etc. In each instance, the parasites may measure from a few millimetres up to a centimetre or so (0.4in) in size, and are thus usually visible to the naked eye. At x10 to x100 magnification, they can be seen to have an oral sucker (around the mouth) and ventral body sucker, both of which are used for attaching the parasite to the host. Although they have a mouth and gut, these flatworms also absorb nutrients directly through the skin, or tegument.

Well-known examples of digenetic flukes include the sheep liver fluke (*Fasciola hepatica*) and *Schistosoma*, which causes bilharzia in man. Because of their complex life cycles, the digenetic flukes that affect fish, such as those that cause yellow grub (*Clinostomum*), black spot (*Posthodiplostomum*) and eye fluke (*Diplostomum*), rarely warrant attention in the aquarium or garden pond.

Below: A larval parasite from a black spot on fish skin. Actual size 1mm(0.04in). This will mature in the gut of a fish-eating bird.

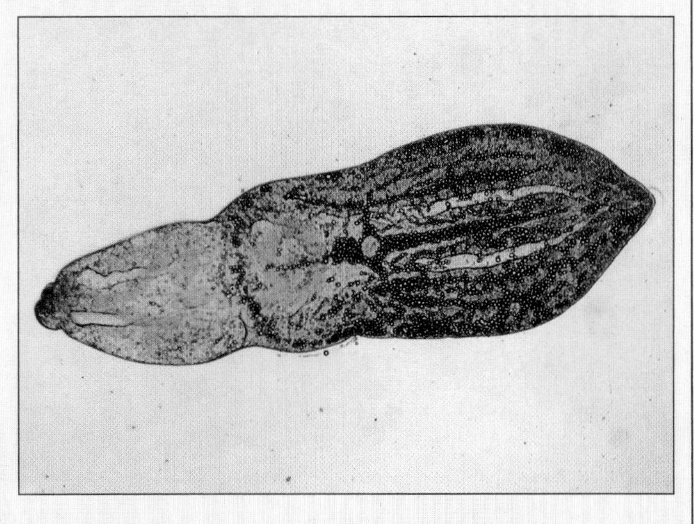

CHAPTER 7

A GUIDE TO TREATMENT

Just as the number of diseases which may affect fish is very large, so the number of possible chemicals and drugs which may be used as treatments is also large. In this book, we have attempted to keep the number of chemicals and drugs to a minimum, at the same time as avoiding any possible loss of effectiveness in treating the condition successfully.

Although some proprietary brands of remedy are based on easily available chemicals, most aquarists will experience problems in making up or obtaining relatively small volumes of the necessary stock solutions. In addition, the cost of such stock solutions is, in some instances, unlikely to be low, especially if only large amounts of the required chemicals are available. Furthermore, proprietary brands of remedy backed by a proven record and with a background of scientific research and development are very safe to use and will reliably treat the diseases which they were developed to combat. When using home-made treatments, it is very easy to overdose the fish, especially if you have made a miscalculation preparing the stock solution. On the other hand, it is almost impossible to overdose branded treatments if you follow the manufacturer's instructions.

Of course, accurate diagnosis (see Chapter 5) goes hand-in-hand with prompt and correct treatment. Having diagnosed the nature of the problem, it is vital to add the correct dose of any remedy to the pond or aquarium. While overdosing might kill the fish, underdosing is unlikely to effect a cure, and may precipitate other problems.

Calculating dose rates

For measuring out large volumes of water in which to mix up remedies, either use a bucket with calibrations on the side, such as those sold in home brew shops, or calibrate a bucket by filling it from a smaller container of known size and marking the side with a waterproof pen. It is sometimes useful to calibrate the treatment tank in a similar fashion as well.

Where necessary, measure small volumes of fluid using plastic disposable syringes or 5ml medicine spoons (available from pharmacies) or a measuring cylinder (available from a pharmacy, home brew shop or photographic shop). Eye-droppers (for counting

Left: These cichlids from Lake Tanganyika (*Lamprologus brichardi*) thrive in hard, alkaline water. They are very sensitive to changes in water quality and thus need careful management when being treated.

drops) are also available from pharmacies. It is fortunate that most good-quality branded treatments are supplied in clearly graduated containers that make correct dosing so much easier.

Before adding the chemical to a tank, mix or dissolve the required amount in a jar full of water from the tank and then mix it thoroughly throughout the bulk of the aquarium water. When adding treatments to a pond, mix or dissolve the remedy in one or two watering cans full of pond water, and then sprinkle it all over the pond surface. (Make sure that the watering can has not been used to spread garden pesticides or other chemicals.) It is probably unwise to use highly coloured plastic containers or hoses, since some of these may give off poisonous resins. Also, never mix remedies in, or treat fish in, galvanized metal containers.

The treatment tank

In some situations, it is advisable (or even essential) to treat fish in a separate isolation, or treatment, tank. This can be set up along similar lines to a quarantine tank (see Chapter 4, pages 72-73), and the quarantine tank will easily double as a treatment tank.

During the treatment period, take care to avoid overcrowding and be sure to provide adequate aeration. It is also vital to disinfect the tank and all its equipment thoroughly after each disease treatment has been completed, and to use a separate set of equipment, such as nets, scrapers, siphon, hose, bucket, etc., for the treatment tank. As described on page 198, bleach is a useful disinfectant, but be sure to use it carefully since it is corrosive and toxic to fish and plants.

Calculating volumes

The volume of an aquarium can be calculated thus:

length x width x water depth (all in cm) = volume in litres ÷ 1000

To convert litres to US gallons, multiply by 0.26; to convert litres to Imp. gallons, multiply by 0.22. If significant amounts of gravel, rocks and other decorations are present, reduce this figure by 10-20 percent.

Calculate the volume of a square or rectangular pond in a similar fashion, but working in metres. Thus:

Length x width x average depth (all in metres) x 1000 = volume in litres

Calculating the volume of a circular pond is also quite straightforward:

Radius of pond x radius of pond x 3.14 x average depth (all in metres) x 1000 = volume in litres.

Calculating the volume of irregularly shaped ponds is more difficult. Fortunately, many homes are nowadays fitted with water meters making it easy to calculate the amount of water volume required to fill any shape of pond. Alternative, when first filling a pond with a hose pipe, use a stopwatch to record the time taken to fill, say, a 10 litre bucket. It is then a simple matter to estimate the volume of the pond based on the time taken to fill it at the same flow rate. Another option is to purchase an inexpensive flow-meter from a garden centre.

Using stock solutions

Desired concentration in mg/litre
0.01
0.1
1.0
10.0

The table indicates the quantity of each stock solution (in ml) needed to achieve the desired concentration in 10 litres of water to be treated.

- 0.1 percent stock solution = 1gm chemical in 1 litre of water
- 1.0 percent stock solution = 10gm chemical in 1 litre of water
- 10 percent stock solution = 100gm chemical in 1 litre of water

Thus, stronger stock solutions are better suited where a final desired concentration is high *or* where the volume of water to be treated is large. Obviously, overdosing is easier when using stronger stock solutions.

Stock solution		
0.1 percent	1.0 percent	10.0 percent
0.1ml	0.01ml	0.001ml
1.0ml	0.1ml	0.01ml
10.0ml	1.0ml	0.1ml
100.0ml	10.0ml	1.0ml

General hints on treatment

When dealing with expensive and/or delicate fish species, it is often a good idea to try the treatment out on one or two individuals before treating a whole batch. Young or chronically diseased fish may also be more sensitive to some treatments.

Local water conditions, particularly pH and hardness levels, may influence the toxicity of certain chemicals to fish. However, most reliable, branded remedies take this into account in their formulation, which is a further reason for favouring proprietary brands of remedy over home-made recipes.

During treatment, especially the first 30-60 minutes, observe the fish for any signs of distress. Should a fish appear ill-at-ease at any time, remove it immediately into clean, aerated water at a similar temperature.

Filtration over activated carbon, excessive amounts of organic matter, protein skimming and water treatment involving ozone all reduce the effectiveness of many treatments. Certain treatments have an adverse effect on the useful bacterial flora of filters and should not be used in tanks which are heavily stocked and rely to a large extent on biological filtration. Some treatments may also be toxic to plants, and marine invertebrates are very sensitive to a number of commonly used treatments. Always investigate these points before using treatments of uncertain potency.

Depending upon the disease in question, the fish should show some signs of improvement within a week of the first treatment. If necessary, carry out partial water changes and apply a further course of treatment. Most proprietary brands of remedy are supplied with clear instructions about repeated and/or prolonged treatments.

Occasionally, if the first remedy does not effect a cure, it may be necessary to use a different chemical or drug. However, avoid mixing different chemicals, and the successive use of a vast array of chemicals. Where it is imperative to change to a different remedy, making a 50-75 percent water change, along with filtration over activated carbon for 12-24 hours, should be sufficient to remove most of the active ingredients of the original remedy.

Sensible precautions

Remember to treat all chemicals and disease treatments with respect, as they may be toxic to humans as well as aquatic animals. As a precaution, wear protective gloves – especially when handling powder or concentrated stock solutions – and never allow contact with eyes, skin and mouth. Store chemical remedies well out of reach of children and pets, and if in doubt about any disease or treatment, refer to a local veterinarian.

Treating pests and diseases

A variety of proprietary treatments, such as those effective against finrot, fungus, white spot, velvet, etc., are available from aquarium shops, along with a number of broad-spectrum aquarium antibacterials and antiparasite treatments. These are often based on easily available chemicals, some of which will be referred to below. If used correctly, such proprietary treatments can be very effective and it is important to emphasize that in many instances it is far safer to use a reliable branded remedy than to produce and use a home-made recipe. Always follow the manufacturer's instructions and choose a brand with a proven record and a background in scientific research.

The treatments featured in this section are presented under a number of broad headings. Each treatment is described, with information on dose rates or concentrations, plus details of any possible drawbacks. Additional information and summaries of the concentrations recommended to treat various health problems effectively appear in accompanying panels and tables.

TREATMENTS AGAINST ALGAE

Various chemical algicides can be used to combat algae in the aquarium or garden pond, but it is important not to rely solely on chemical methods for the control of this type of problem. When an algal problem occurs, always investigate what management procedures may be used to bring the situation under control (see page 71 in Chapter 4 and page 134 in Chapter 6). Ideally, algicides should only be used as a last resort.

When chemical treatment is necessary, you will find it most convenient to use one of the proprietary algicides available from aquarium shops.

One herbicide that has been used with reasonable success by koi-keepers is terbutryne (e.g. Clarosan, Ciba-Geigy). It is normally used concentrated 5 to 10mg/litre, and one application normally brings troublesome blanket weed (a filamentous algae) under control within 14 days. Clarosan, is, in fact, terbutryne formulated as slow-release granules, which require several days to have their maximum effect.

Barley straw has gained popularity as a natural method for controlling blanket weed in ponds. Compact pads of barley straw are available specifically for pond use.

Some success has also been achieved using high-rate mechanical filters and/or ultraviolet (UV) sterilizers to control 'green water' problems in garden ponds. To be successful, such mechanical filters must not only be capable of removing tiny suspended algal cells and easily serviced to prevent clogging, but also pass water through their medium or media faster than the algae are multiplying. The use of UV sterilizers is discussed on pages 199-200.

Caution

● Some chemical algicides can harm fish if they are not used strictly according to the manufacturer's instructions.

● Aquatic plants may be harmed, even at the concentrations used to control algae. If in doubt, consult the manufacturer. As a result, chemicals such as terbutryne are normally used to control algae in *unplanted* koi ponds.

● The sudden die-back of a large amount of algae following the use of an algicide can adversely affect water quality. If possible, physically remove large amounts of algae from the pond or aquarium before treatment.

TREATMENTS AGAINST BACTERIA AND FUNGI

A range of chemicals and drugs can be used to treat bacterial and fungal diseases of fish, although their ease of availability may vary from country to country. Fortunately, proprietary treatments are available from aquarium shops to treat many of the common problems, such as fish and egg fungus, finrot, cotton-wool disease, etc., but veterinary assistance may be required to obtain and/or formulate some of the treatments listed below. Your local aquarium shop should be able to tell you if any proprietary treatments contain similar chemicals or drugs.

Right: Angelfish eggs in water treated with methylene blue to provide protection against egg fungus. The shade gives a guide to the concentration required.

Antiseptics and open wounds

Fish that are physically damaged or suffering from the effects of ulcer disease, anchor worm, etc., may have lesions on their body that should be treated with a topical antiseptic – in addition to any remedy added to the water. Two such antiseptics are mentioned here.

Mercurochrome is available from most pharmacists and veterinarians. The required stock solution is 2 percent (in water). Used neat (or diluted 1:9 with clean tapwater), it may be painted onto wounds, fungal infections and the like. Several treatments on alternate days may be needed.

Caution

● Handle the stock solution with care, avoiding contact with skin, eyes and mouth.

Veterinary iodine-based antiseptics may also be used to

treat local lesions on fish. These are available in spray or liquid form, and can be applied in a similar fashion to that described for mercurochrome. They are usually applied without dilution.

Once annointed with antiseptic, open wounds may be smeared with vaseline or a waterproof wound dressing, which not only reduces the likelihood of infection from the water but also limits the extent of osmotic problems via the exposed flesh. During such a procedure, handle the fish very gently, wrapping it in a soft damp cloth, and keeping it out of the water for the minimum period possible.

Below: Applying mercurochrome as an antiseptic to damaged koi scales. Use a cotton swab and take care with this preparation.

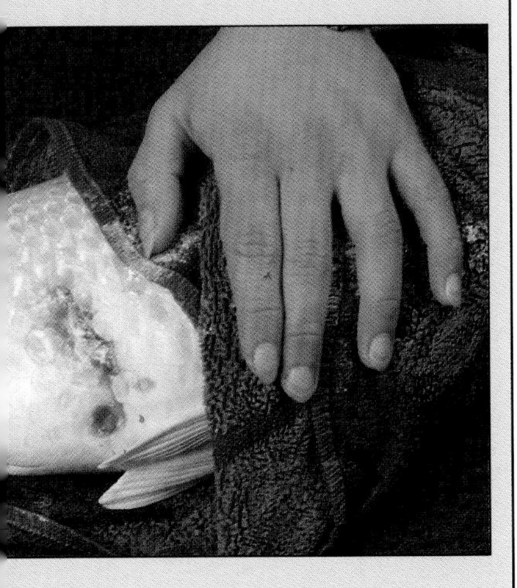

Methylene blue

This is available from aquarium shops, usually in a 1 or 2 percent solution. However, its use in treating fish diseases is now somewhat outdated. At around 2mg/litre (2ml of a 1 percent solution per 10 litres) it has been used to treat egg fungus, and also some external protozoan infestations of fish. Add the chemical to the tank water and leave it as a continuous bath for several days.

Caution

- Methylene blue is easily deactivated by organic matter.
- It may adversely affect plants.
- It may adversely affect the helpful bacteria in a biological filter, and this brings about rises in ammonia and/or nitrite.
- It may discolour aquarium equipment, including silicone sealant.

Phenoxyethanol (2-phenoxyethanol)

This oily liquid can be used to treat diseases such as finrot and cotton-wool disease. It is normally used at a concentration of 100mg/litre (50ml of a 2 percent solution per 10 litres of water). Add this to the water and leave it for at least seven days. (Phenoxyethanol is also used as a fish anaesthetic – see page 79).

Caution

- Phenoxyethanol may 'leach' other chemicals from activated carbon. (Note: For this and other reasons, suspend filtration over activated carbon during *all* disease treatments.)

Antibiotics and similar drugs

These are controlled by the veterinary profession in many countries and, consequently, it may be necessary to seek veterinary advice before undertaking this type of treatment. In the USA, however, some antibiotics are directly available in most aquarium stores. When dealing with fish, there are three methods for administering antibiotics and similar drugs: by injection, by mixing the drug with the food, or by adding it to the water. The first two options are usually preferred for treating systemic bacterial problems, such as ulcer disease, although some drugs may be effective when added to the water to treat external infections, such as finrot and cotton-wool disease. Very often, adding the drug to the water may be the only method available to fish hobbyists.

The use of oxytetracyline hydrochloride, an antibiotic commonly used to treat fish, is described in detail below, although other similar drugs are mentioned where pertinent. A local veterinarian may be able to suggest one or two alternatives, depending on the disease.

Here we look in more detail at the three alternative methods of administering antibiotics and similar drugs.

By injection. This should only be carried out by trained and qualified personnel. Administering antibiotics to fish by injection is only applicable to relatively large, easily handled species. An antibiotic preparation manufactured specially for injection must be used, since the injection of other forms of antibiotics could be dangerous for the fish. A dose rate of 10-20mg of oxytetracyline hydrochloride per kilogram of fish (when injected into the

Aquarium antibacterials and anti-fungals

Chemical	Dose rate/ concentration	Duration/frequency	To treat
Chlortetracycline	10-20mg/litre	Continuous bath for up to 5 days; may need repeating	Finrot, cotton-wool disease, systemic bacterial infections (e.g. ulcer disease)
Furazolidone	50-75mg/ kg fish	With food; each day for 7-10 days	Systemic bacterial infections (e.g. ulcer disease)
	20mg/litre	Continuous bath for up to 5 days; may need repeating	Symptoms associated with neon tetra disease
Kanamycin sulphate	12-13mg/litre	Continuous bath; add every day for at least 5 days; make partial water changes between treatments	Wasting disease
Methylene blue	2mg/litre	Continuous bath for several days; may need repeating	Egg fungus, external protozoans
Nifurpirinol	0.1-0.2mg/ litre	Continuous bath for up to 5 days; may need repeating	Finrot, cotton-wool disease, systemic bacterial infections (e.g. ulcer disease)
	Up to 2mg/ litre	5-10 minute bath; may need repeating	Finrot, cotton-wool disease, systemic bacterial infections (e.g. ulcer disease)
Oxolinic acid	10mg/kg fish	With food; each day for 10 days	Systemic bacterial infections (e.g. ulcer disease)
Oxytetracycline hydrochloride	10-20mg/ kg fish	Injection; may need repeating	Systemic bacterial infections (e.g. ulcer disease)
	60-75mg/ kg fish	With food; each day for 7-14 days	Systemic bacterial infections (e.g. ulcer disease)
	20-100mg/ litre	Continuous bath for up to 5 days; may need repeating	Finrot, cotton-wool disease, systemic bacterial infections (e.g. ulcer disease)
Phenoxyethanol	100mg/litre	Continuous bath for at least 7 days	Cotton-wool disease, finrot
Potentiated sulphonamide (e.g. Tribrissen)	30-60mg/ kg fish	With food; each day for 5-7 days	Systemic bacterial infections (e.g. ulcer disease)
Proprietary treatments	Use as directed	As directed	Fungus, finrot, cotton-wool disease, etc
Sulphadimidine	200mg/ kg fish	With food; each day for up to 14 days	Systemic bacterial infections (e.g. ulcer disease)
Sulphafurazole	0.2mg/ gm fish	Injection; may need repeating	Wasting disease
plus doxycycline hydrochloride *or* minocycline hydrochloride	0.005mg/ gm fish		

peritoneal cavity) has been found effective in treating certain bacterial diseases of fish. However, antibiotics may also be injected into the musculature of fish and, although a certain amount of the drug may be squeezed out by muscular contractions, this latter route is probably the safest.

After an injection, keep the affected fish in an isolation tank for several days, and if there are no signs of improvement, seek further advice. When all signs of the disease have disappeared, reintroduce the fish into the set-up aquarium or pond.

For treating 'wasting disease' (see page 162), reports suggest that the following drugs have been used with some success on fish: sulphafurazole at 0.2mg/gm fish, *plus* doxycycline hydrochloride or minocycline hydrochloride at 0.005mg/gm fish. Unfortunately, such drugs are expensive and difficult to administer to small fish via the required intramuscular injection. Where fish suffering from wasting disease cannot be treated by injection, adding oxytetracycline hydrochloride or kanamycin sulphate to the water may be worthwhile (see below). However, for chronic cases of wasting disease, repeated treatments over at least a 10- to 14-day period may be needed.

By mixing with food. This is difficult on a small scale, although medicated flaked and/or pelleted foods are available in Europe and North America. (Consult your aquarium shop or a local veterinarian for information.) Unfortunately, diseased fish often stop feeding, which can limit the use of this type of treatment. Larger fish, such as koi and large goldfish, may be fed medicated food by making a slurry of a little cooking oil and antibiotic, which can then be mixed with a pelleted food. A dose rate of 60-75 mg of oxytetracycline hydrochloride per kg of fish, fed every day for 7-14 days, has been found effective in treating some bacterial diseases of fish. Other similar drugs which have been used in this fashion include: furazolidone/nitrofurazone at a dose rate of 50-75 mg/kg fish per day for 7-10 days; sulphadimidine at a dose rate of 200mg/kg fish per day for up to 14 days; potentiated sulphonamide, such as Tribrissen (a mixture of trimethoprim and sulphadiazine) at 30-60 mg/kg fish per day for 5-7 days; oxolinic acid at a dose rate of 10mg/kg fish per day for 10 days.

By addition to the water. To treat stubborn cases of finrot and cotton-wool disease, antibiotics and similar drugs may be added to the water. Some success has also been experienced in treating more systemic bacterial problems in this way. When using oxytetracycline hydrochloride (soluble powder form), it is normal to use a concentration of 20mg/litre as a five-day bath in an isolation tank. Dissolve the antibiotic in a little water from the tank and then mixed it thoroughly with the remaining bulk of the water. Leave the fish in the treated isolation tank for at least five days. If at any stage the fish become noticeably distressed, carry out a partial water change. If there are no signs of improvement after five days, make a water change and repeat the above dosage (watching for signs of distress in the fish), or contact your local veterinarian for advice. Unfortunately, this method of treatment has a drawback in that hard water may interfere with the antibiotic in solution or prevent

Antibacterial compounds

This is a general term describing any compound which is active against bacteria. Some relatively simple compounds, such as formalin, have an antibacterial action. Complex compounds, such as antibiotics, are also generally antibacterials. Antibacterials are used in the aquarium to treat disease or to sterilize equipment. It is important to prevent these compounds harming beneficial bacteria, particularly in filters.

Antibiotics

The term 'antibiotic' is frequently misused. Strictly speaking, antibiotics are substances produced during the growth of micro-organisms that in low concentration destroy or inhibit the growth of certain other micro-organisms. The term has now been extended to include chemically related or derived substances. Some antibiotics have no activity against micro-organisms, but have other uses.

Below: Some antibiotics can be delivered via the food, provided the fish haven't lost their appetites.

its uptake by the fish. Higher dose rates (up to five times the above) may be required under these circumstances.

Other antibiotics which have been used in this way include: chlortetracycline at 10-20 mg/litre; furazolidone/nitrofurazone at 20mg/litre, specifically to treat the symptoms of neon tetra disease; kanamycin sulphate at 12-13mg/litre per day for at least five days, specifically to treat fish with wasting disease.

Nifurpirinol (Furanace) is a chemical that has showed great promise for treating a range of diseases of aquarium fish. However, it has rather limited availability in certain parts of the world. It is often used at a concentration of 0.1-0.2mg/litre for between three and five days. Much higher concentrations (up to 2mg/litre) have been used in the form of very short baths, i.e. 5-10 minutes, in an isolation tank.

Caution

● Antibacterial chemicals may affect the helpful bacterial flora in biological filters (see the table opposite). Therefore, unless you know it to be safe, do not add such treatments to the water of the set-up aquarium or pond, but use them in a separate treatment tank as shown on pages 72-73.

● Using certain antibiotics and drugs at too low a dose rate, or for inadequate periods of time, or to treat diseases for which they are not recommended, may result in resistant strains of disease organisms. Thus, consult a local veterinarian before using such chemicals to treat fish and always use the full, recommended dose rate for the prescribed length of time.

TREATMENTS AGAINST EXTERNAL PARASITES

An enormous range of chemicals has been used to treat external parasite infestations of fish. Many of these are easily available from pharmacists or veterinarians, or, in the form of tried and tested proprietary treatments, from aquarium shops. We have already considered the benefits of proprietary treatments over home-made recipes, but if you want to make up your own remedies a local pharmacist or veterinarian should be able to help you prepare simple stock solutions and calculate therapeutic concentrations. Here, we consider a number of chemicals effective against external parasites; good aquarium shops will be able to say whether they are available as easy-to-use proprietary treatments.

Formalin

This is a solution of formaldehyde gas in water, normally available as a 37-40 percent solution from pharmacists. It can be used to treat a range of external fish pathogens, especially skin and gill flukes, sliminess of the skin and marine white spot. In freshwater aquariums, a concentration of 15-25mg/litre (0.15-0.25ml or 4 drops commercial strength formalin per 10 litres) as a continuous bath for several days can be used to treat skin flukes, gill flukes and sliminess of the skin. As a result of its toxicity to marine invertebrates and, perhaps, the bacteria responsible for biological

Effect of selected disease treatments on nitrification

Antibacterial	Concentration
Copper	0.3-0.5mg/litre
Chlortetracycline	10mg/litre
Formalin	15-25mg/litre
Malachite green	0.1-0.5mg/litre
Methylene blue	1-8mg/litre
Nifurpirinol	0.1-1.0mg/litre
Oxytetracycline	50mg/litre
Sulphamerazine	50mg/litre

Note: This table summarizes a number of recent studies. Apparent lack of effect or non-inclusion should not be taken as a definitive result.

Right: Heavy infestations of monogenean flukes, such as this skin fluke, may seriously harm fish if not promptly treated.

Comments

Known to affect nitrogen cycle in salt water; no apparent effect in fresh water

Known to affect nitrogen cycle in fresh water. Few data on salt water.

Variable effects on nitrogen cycle noticed in fresh water. May affect nitrification in salt water.

May affect nitrogen cycle in fresh water. Few data on salt water.

Can cause severe and prolonged interruption to nitrogen cycle in fresh and salt water.

No apparent effect in fresh water; may affect nitrification in salt water.

Known to affect nitrogen cycle in fresh water. Few data on salt water.

No apparent effect on nitrification in fresh water. Few data on salt water.

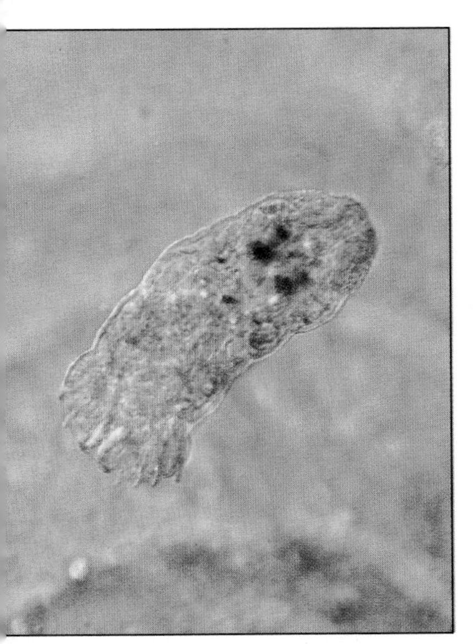

filtration in seawater systems, its use to treat marine fish is often restricted to a separate treatment tank. Here a 30- to 60-minute bath in 200mg/litre (2ml commercial strength formalin per 10 litres) can be used to control sliminess of the skin, marine white spot, and skin and gill flukes. Such a treatment may need repeating after a few days and is best coupled with a copper or similar antiparasite treatment to the infested display tank.

Caution

- Formalin is a very unpleasant chemical and must be kept away from eyes, mouth and skin. Only use it in a well-ventilated area and do not inhale the vapour.
- Most fish fry and some fish, such as marine surgeonfishes, react badly to formalin, especially at higher concentrations. Always observe the fish during treatment and remove them to clean, well-aerated water at the first sign of distress.
- Formalin removes oxygen from the water and can also damage the gills. Aerate the water thoroughly during formalin treatment and take special care with fish with obvious gill problems.
- Some plants and marine inverts are very sensitive to formalin.
- On storage, a very toxic white precipitate develops in formalin. Do not use formalin from a bottle containing this white precipitate. Always store away from light to prevent the precipitate developing.
- Formalin is more toxic in soft water; use with special care when treating freshwater fish in soft water.

Malachite green

This green dye is active against a wide range of external parasites and pathogens (including fungi, bacteria, skin and gill flukes), but it is most useful for treating external protozoan parasites in freshwater systems. It is frequently used at a concentration of 0.1-0.2mg/litre (0.1-0.2ml of 1 percent stock solution per 10 litres) as a continuous bath for several days. Two or three doses are usually added over a 7- to 10-day period. In set-up aquariums and ponds, there is no need for a partial water change between treatments. In the more sterile conditions of a quarantine tank, a 25 percent water change before each repeat treatment is recommended. On occasion, several one-hour baths in 1-2mg/litre malachite green (1-2ml of 1 percent stock solution per 10 litres) over a 7- to 10-day period may be useful as an alternative. Concerns that this dye, especially in neat powder form, may be harmful to human health have reduced its popularity as a disease remedy.

Caution

- Some fish, such as 'scaleless' catfish and some characins, do not tolerate malachite green very well.
- Malachite green is easily deactivated by organic matter.
- It is wise not to handle the neat powder form of this chemical, due to the risk of inhalation. Handle concentrated stock solutions with care, avoiding contact with skin, eyes and mouth.

Malachite green with formalin

This is a useful treatment which combines the therapeutic effects of both chemicals. It is prepared by dissolving 3.7gm of malachite green into one litre of fresh 37-40 percent commercial strength formalin. Applied at 0.25ml per 10 litres, this gives a final concentration of 0.1mg/litre malachite green and 25mg/litre formalin. This combination is particularly useful for treating sliminess of the skin in freshwater systems. In a pond or set-up aquarium, apply two or three doses over a 7- to 14-day period; there is no need to make a water change between these treatments. However, in a quarantine or treatment tank, make a 25-50 percent water change before each repeat treatment.

Together or separately, formalin and malachite green are common constituents of many proprietary fish disease treatments.

Caution

● See *Formalin* and *Malachite green* above.

Copper

Copper is a relatively common treatment for diseases such as coral fish disease, white spot and sliminess of the skin (all in sea water), as well as some of the similar diseases that occur in fresh water. Proprietary copper treatments are readily available from aquarium shops, although a suitable stock solution can be prepared by dissolving 4.5gm of copper sulphate and 3gm of citric acid in 1 litre of water. Using this solution at a rate of 1.3ml per 10 litres of water gives a final concentration of 0.15mg/litre copper. Concentrations up to 0.3mg/litre are used in the control of fish diseases. However, copper is readily absorbed by organic matter and also by calcareous material such as coral, tufa rock, cockleshell, etc., and so it is advisable to check theoretical concentrations with actual concentrations using a test kit. Since most copper treatments depend on therapeutic levels being maintained for up to four weeks, you will need to make regular additions of copper (daily or every other day) and monitor the copper levels regularly to ensure that effective, but not hazardous, levels are maintained.

Caution

● Some fish are very sensitive to copper; true sharks, for example, may be intolerant of normal therapeutic levels. At any rate, never exceed a copper concentration of 0.4mg/litre for any species.

● Invertebrates, some algae and plants, and the helpful bacteria in a biological filter are also sensitive to copper. Therefore, do not use copper in a tank containing invertebrates and use it with care in a tank that relies heavily on biological filtration.

● Since copper is readily absorbed by organic substances and calcareous materials, higher copper levels may be obtained per unit treatment in 'sterile' treatment tanks compared with set-up aquariums. Monitor effective levels with a test kit.

● Copper can leach out from calcareous rocks that have absorbed it, perhaps harming invertebrates that were not resident in the tank during treatment. Filtration over activated carbon may help to resolve this problem.

● Proprietary tapwater conditioners may reduce copper activity.

● When used in fresh water, copper can be noticeably more toxic to fish in soft water than in hard water; use it with special care under such conditions.

Treatments effective against external parasites

Chemical	Dose rate/ concentration
Copper	0.15-0.3mg/ litre
Formalin (37-40 percent)	15-25mg/litre
	200mg/litre
Freshwater dip	(Check pH and temperature and adjust if necessary)
Malachite green	0.1-0.2mg/ litre
	1-2mg/litre
Malachite green with formalin	0.1mg/litre with 25mg/ litre
Metriphonate	0.25-0.4mg/ litre
	Up to 1.0mg/ litre
Potassium permanganate	10-20mg/litre
	10mg/litre
Proprietary treatments	Use as directed
Salt	See table

Duration/ frequency	To treat
Continuous bath up to 4 weeks; monitor with test kit	Skin and gill protozoans, flukes, etc.; fresh/sea water
Continuous bath for several days	Skin and gill flukes; sliminess of the skin; fresh
30-60 minute bath; may need repeating	Skin and gill flukes; sliminess of skin; marine white spot; fresh/sea water
Short bath for 2-10 minutes; may need repeating	Coral fish disease; white spot; sliminess of skin; sea water
Continuous bath for several days; may need repeat doses	Fungus; white spot; sliminess of the skin; velvet; fresh
One hour bath; may need repeating	Fungus; white spot; sliminess of the skin; velvet; fresh
Continuous bath for several days; may need repeating	Sliminess of the skin; other external protozoans and flukes; fresh water
Continuous bath for 7-10 days; may need repeating	Skin and gill flukes; parasitic crustaceans; leeches; fresh/ sea water
Continuous bath for 7-10 days; may need repeating; use with special caution	Stubborn leeches and marine black spot disease
Short bath for 30 mins.; may need repeating	Active against fish lice
Short bath for 10 minutes	General plant disinfectant
As directed	Most common infestations

Freshwater dips

These can be used to treat a number of external parasites of marine fish, especially sliminess of the skin, coral fish disease and marine white spot. Such a treatment is not necessarily effective on its own, but is often used in conjunction with other remedies, such as copper. To make a freshwater dip, partially fill a small aquarium or bucket and bring the water up to the required temperature with a little boiling water from a kettle. Use a proprietary conditioner where necessary to remove chlorine from tapwater. Using a commercial buffer solution or a little baking soda, adjust the pH value of the fresh water to within 0.2-0.3 units of that in the marine tank. Provide gentle aeration. Lightly restrain the fish to be treated within a spacious, soft hand net in the fresh water for two to ten minutes. Coral fish disease and white spot need a short bath (two to three minutes), while the flukes involved in sliminess of the skin may need five to ten minutes. Be sure to return the fish to sea water if any signs of distress occur. Such a treatment may need repeating two or three times over a 7- to 10- day period.

Caution

- Some marine fish are particularly sensitive to freshwater dips.

Salt (sodium chloride)

Salt has a number of important uses in freshwater aquariums and ponds, but it is not a universal panacea. Although sometimes recommended as a routine 'tonic' for freshwater fish, salt should only be added to an aquarium when necessary, notably for treating certain disease conditions or for eliminating some pests and parasites. At relatively low levels (up to 0.3 percent), it can be used to help reduce the stress associated with physical damage and/or high nitrite levels. Exposure to 0.3-0.5 percent salt for three to five days can control *Hydra*, and a continuous bath in a 1 percent salt solution is a useful supportive treatment for coldwater fish suffering from ulcer disease. A 15- to 30-minute bath in a 2-3 percent salt solution will remove leeches from pond fish, or at least cause them to loosen their hold so that you can remove them easily with forceps. For the above situations, completely dissolve the dry salt in a small amount of water from the aquarium or pond and then disperse it throughout the water to be treated. When necessary, gradually remove the salt from a set-up aquarium or pond by carrying out partial water changes with fresh water.

Caution

- Use cooking salt or aquarium salt, but *not* table salt, which has other additives.
- Some fish, such as certain 'softwater' species and some catfishes, are more sensitive to salt that others.
- When using continuous baths of 0.5-1 percent, increase the salt concentration in gradual steps over one to two days to prevent osmotic shock to the fish. Above 1 percent, even short-term baths can be stressful, especially if prolonged. Always remove distressed fish to clean water promptly.
- Plants may be affected by prolonged exposure to levels of salt in excess of 0.5 percent.

Potassium permanganate

This purple crystalline compound has limited use in the treatment of pests and diseases of fish. However, it can be used to treat fish suffering from lice, where the fish is susceptible to treatment with organophosphorus insecticide compounds (see below). Lice infested fish are usually given a 30-minute bath in 10-20mg/litre potassium permanganate (10-20ml of a 1 percent solution in 10 litres of water). In addition, a concentration of 10mg/litre can be used as a 10-minute bath as a general disinfectant for aquarium plants (see also Chapter 4, page 73).

Caution

- Do not add crystals of potassium permanganate directly to the water containing the fish; use a stock solution and disperse the required amounts throughout the water to be treated.
- Always store the stock solution in the dark, as it decomposes in bright light.
- Potassium permanganate is easily deactivated by organic matter.
- Some fish are sensitive to potassium permanganate, especially in alkaline water conditions.

Organophosphorous insecticidal compounds

These compounds, particularly metriphonate (trichlorfon), have been used to treat a number of external fish parasites, notably skin and gill flukes, leeches and crustacean parasites in both fresh water and sea water. In recent years, concerns over possible neurotoxic effects on human health and the environment have led to these chemicals being restricted or prohibited in certain countries. Consult your aquarium store or veterinarian for advice on restrictions and availability.

Metriphonate is normally used at 0.25-0.4mg/litre as a continuous bath for seven to ten days. Since the chemical degrades very quickly in warm, alkaline water, it is normal to add additional amounts of metriphonate to maintain therapeutic levels for the duration of the treatment period. In fact, only slightly more than half of an original single treatment will be left after 24-48 hours,and almost all of it will have been degraded after seven to ten days in a tropical marine aquarium. Proprietary aquarium and pond formulations take this into account and are supplied with clear instructions for use.

Concentrations up to 1.0mg/litre metriphonate have been used to control stubborn outbreaks of leeches and marine black spot caused by a turbellarian flatworm. Such treatments should be applied with special caution.

At the concentrations normally used to treat fish, plants are not affected and there appear to be no adverse effects on nitrification. Similarly, zooplankton and other invertebrates in garden ponds, which may be initially affected, soon reappear after the treatment has subsided.

Caution

- Metriphonate is an insecticide which acts on the nervous system. Handle it with great care, particularly when working with the dry

Using salt in the freshwater aquarium/pond

Recommended strength (percent)	gms/ litre
0.01 percent	0.1gm
0.1 per cent	1gm
0.3 percent	3gm
0.3-0.5 percent	3-5gm
1 percent	10gm
2-3 percent	20-30gm

NB A 3 percent salt solution is approximately equivalent to full-strength sea water.

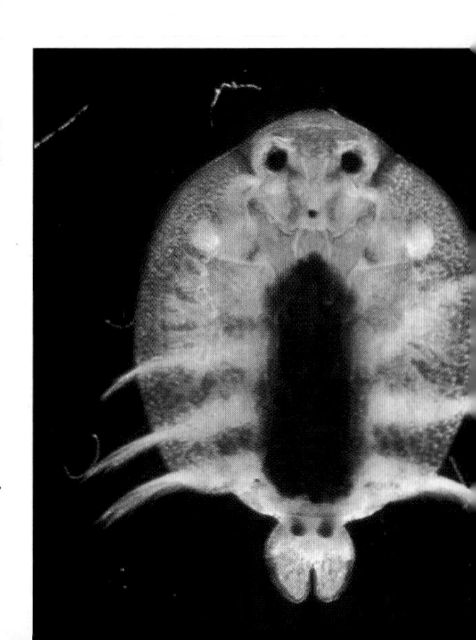

Use
To reduce nitrate toxicity in fresh water. Most freshwater fish are tolerant of this low level of salt. Use as a long term bath until the nitrite concentration falls to a safe level.
Supportive treatment for fish showing signs of physical damage. Use as a continuous bath. Used as an aquarium additive (continuous bath) for certain species of fish (e.g. mollies) that benefit from slightly brackish conditions.
Supportive treatment for physical damage. Use as a continuous bath.
To control *Hydra*. As a continuous bath for 5-7 days
Supportive treatment for coldwater fish with ulcer disease. Acclimate gradually, then use as a continuous bath
To remove leeches from pond fish. Short bath for 15-30 minutes only

Below: A pair of fish lice (*Argulus* sp). The male (right) is smaller than the female. Safer alternatives to organophosphorus compounds are being sought to treat fish lice and other crustacean parasites.

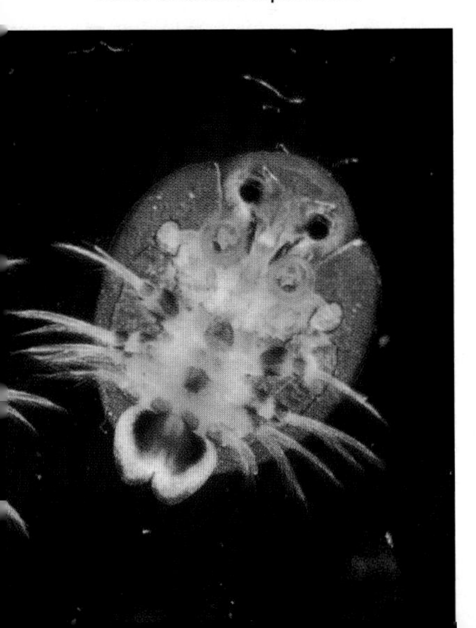

formulations or concentrated solutions. Avoid contact with the skin, eyes and mouth, and handle the dry powder in a well-ventilated area.

- Some fish, such as certain characins (including piranha), orfe, rudd, and also marine surgeonfish and invertebrates, (such as crustaceans and anemones) are very sensitive to metriphonate. Since metriphonate rapidly decomposes in warm, alkaline water, it is usually possible to return such delicate animals to the treated aquarium or pond 10-14 days after the last treatment.
- Metriphonate takes longer to break down in cold water with a pH value below 7.0.
- With time, some parasites may be able to develop resistance to metriphonate.

Ultraviolet irradiation and ozone

These techniques may be used to reduce the levels of microbial disease agents, including small parasites, that are free swimming in water, and reduce the likelihood of their transmission between tanks on a centralized filtration system. (Such methods are discussed in more detail later in this chapter, on page 199.) However, the effectiveness of both methods is linked with parasite size. Hence, small parasites such as *Piscinoodinium* and *Amyloodinium* and the infective stages ('swarmers') of *Ichthyophthirius* and *Cryptocaryon* may be controlled, whereas larger parasites such as flukes etc., do not succumb so easily to the effects of ultraviolet irradiation or ozone.

TREATMENTS AGAINST INTERNAL PARASITES

Fish are host to a startling array of internal parasites, although many of these infestations pass off unnoticed. Since most internal parasites do not appear to cause the major problems associated with external parasites in aquariums and ponds, chemical treatments are few and seldom needed. However, from time to time internal parasites, such as *Hexamita*, fish heartworm (*Sanguinicola*), and intestinal tapeworms and roundworms, may warrant treatment. Here, we consider some of the treatment chemicals involved.

Dimetridazole and metronidazole

These chemicals have been used to treat a variety of infestations in animals and man. Their main use to date, with fish, has been to treat hole-in-the-head disease (*Hexamita*).

Dimetridazole may be used at a final concentration of 5mg/litre in a separate treatment tank. Give this treatment three times at weekly intervals, with a partial water change of about 20 percent before each treatment. It has been suggested that markedly diseased fish should be given a 48-hour bath in 40mg/litre dimetridazole.

Metronidazole can also be used to treat *Hexamita*. One treatment lasting several days in an isolation tank containing 7mg/litre metronidazole is usually sufficient, although this treatment may be repeated every other day for up to three applications – with a 25 percent water change before each treatment.

While proprietary treatments against hole-in-the-head disease may contain these chemicals, veterinarians should also be able to offer assistance in obtaining dimetridazole and metronidazole and in calculating the required concentrations.

Caution

- The above treatments for *Hexamita* have not been tried on a large scale on a wide range of more unusual fish. Take special care in such situations.
- Dimetridazole may inhibit spawning in fish.

Toltrazuril

This drug has been widely used to control coccidian parasites in chickens but has recently proven effective in treating some protozoan diseases of fish, notably microsporidians such as *Pleistophora* which causes neon tetra disease (see page 126). The drug also destroys the parasitic stage of freshwater white spot. Although still undergoing experimental trials, Toltrazuril may become available for treating ornamental fish. Consult your vet regarding current availability of this drug.

Anthelmintics

These are available for the control of certain internal 'worm' infestations, notably tapeworms, roundworm and, perhaps, fish heartworm (*Sanguinicola*). While some of these chemicals may be found in proprietary dog and cat worming tablets or capsules, always seek veterinary assistance before selecting the most useful treatment for fish and starting anthelmintic therapy.

Although some anthelmintics are available in injectable formulations, most of these chemicals are usually administered with the feed. (The preparation of medicated feed is described for oxytetracycline hydrochloride on page 189.) The dose rates of anthelmintic given below are usually mixed with a five-day ration of food (usually pellets) and then fed to all the fish in an infected pond or aquarium over five consecutive days. To ensure adequate control, this procedure should be repeated after 10-14 days.

When dealing with small numbers of large and easily handled fish, it is more effective to treat fish individually using a stomach tube. The dose rate is suspended in a small amount of saline and gently discharged into the stomach of each fish. Once again, to ensure adequate control, repeat this after 10-14 days.

The anthelmintics that have been used to treat internal parasite infestations of fish include:

Niclosamide at 50-100 mg per kg of fish is said to be effective against most intestinal tapeworms, spiny-headed worms and digenetic flukes.

Mebendazole at 25-50mg per kg of fish is active against some intestinal tapeworms in fish, and has been used against skin flukes.

Levamisole has been used at dose rates of up to 200mg/kg body weight to treat a wide variety of nematode infestations in a range of animal species. Application via a stomach tube may be useful when treating fish, although levamisole can also be injected and perhaps

Treatments effective against internal parasites

Chemical	Dose rate/ concentration
Dimetridazole	5mg/litre
	40mg/litre
Furazolidone	50-75mg/ kg fish
	20mg/litre
Levamisole	Up to 200mg/ kg fish
Mebendazole	25-50mg/ kg fish
Metronidazole	7mg/litre
Niclosamide	50-100mg/ kg fish
Piperazine citrate	25mg per 10gm flaked food
Praziquantel	5-100mg/ kg fish
Toltrazuril	20mg/litre

absorbed from the water by fish. A local veterinarian should be able to advise you.

Method	To treat
Continuous bath; may need repeating	*Hexamita* (hole-in-the-head disease)
Short bath for 48 hours	*Hexamita* hole-in-the-head disease)
With feed; every day for 7-10 days	*Hexamita* (especially on trout farms)
Continuous bath for up to 5 days; may need repeating	Symptoms associated with neon tetra disease
Variety of methods possible; seek veterinary advice	Internal nematodes
With feed; over several days; may need repeating	Intestinal tapeworms
Continuous bath; may need repeating	*Hexamita* (hole-in-the-head disease)
With feed; over several days; may need repeating	Intestinal helminths, particularly tapeworms
Feed over 5-10 days; may need repeating	Intestinal roundworms
With feed over several days, or via intramuscular injection; may need repeating	Internal helminths especially tapeworms and (perhaps) fish heartworm (*Sanguinicola*)
Three one-hour baths, given on alternate days	Possibly effective in treating neon tetra disease

Praziquantel at 5-100mg per kg of fish can be used to control a very wide range of internal helminth parasites of fish, especially intestinal tapeworms. Since praziquantel is absorbed into the bloodstream, it may act against helminth parasites elsewhere in the body (including fish heartworm, *Sanguinicola*). An injectable formulation of praziquantel is available, which may be used to treat larger, easily handled fish.

Piperazine citrate has been used to treat livebearers infested with *Camallanus* nematodes. This involves mixing 25mg of piperazine citrate with each 10gm of flaked food, which is then fed to the infested fish over 5-10 days. The flaked food is usually slightly moistened before mixing, although this anthelmintic may be mixed with other foods, too. A repeat course of treatment may be needed after 10-14 days at a dose rate of 50-100mg/kg of body weight.

Caution

- Toxic effects have been observed when using niclosamide in small aquariums or ponds without a good exchange of water.
- The above treatments have seldom been used on a large scale to treat exotic species; take special care when treating unusual fish.

TREATMENTS AGAINST SNAILS

Although they can build up to large numbers in an aquarium or pond, and on occasion act as a source of certain parasite infestations, snails cannot really be regarded as a major threat to the health of ornamental fish. However, while a number of different chemicals can be used to combat snails in an aquarium or pond, such treatments must be nontoxic to fish and plants. A number of proprietary snail treatments are available, which are effective and safe to use. However, be sure to follow the manufacturer's instructions closely, and only use them as a last resort.

Caution

- Water quality problems can suddenly develop in aquariums harbouring large numbers of snails that die following the use of a snail eradicator. Use with care under such circumstances, and remove all dead snails promptly. Some snails, such as the Malayan livebearing snail *(Melanoides)*, may abound, hidden in gravel.

DISINFECTANTS

To improve disease control it may be necessary to disinfect aquariums, ponds, and equipment – and even the water in which the fish swim. Simply hosing or rinsing equipment and fishkeeping facilities in clean water to remove organic debris and then thorough drying is an excellent way of eliminating many fish pathogens.

Bear this in mind when storing seldom-used equipment, or even hand nets, siphon pipes and buckets which are in more frequent use. For more positive disinfection, you can choose from a wide range of easily available disinfectants. We consider two here in detail: household bleach (sodium hypochlorite) and iodophors (iodine-based disinfectants).

Household bleach

Bleach is one of the easiest disinfectants to obtain. Diluted with cold tapwater until a faint but distinct smell of chlorine is still detectable, bleach is effective against a range of fish disease organisms. First rinse the equipment in clean water to remove excessive amounts of debris, and then leave it in contact with the diluted bleach for 30 minutes or so. Rinse the equipment thoroughly in clean water before storage or use in contact with live fish. Wipe the tank and other surfaces with a cloth soaked in diluted bleach, leave them for about 30 minutes and then rinse them thoroughly. Bleach is no longer active when the chlorine smell disappears.

Caution

- Bleach is unpleasant; keep it away from skin, eyes and mouth.
- Bleach is corrosive to metals, and will damage nylon nets.
- Bleach is easily deactivated by organic matter.
- Bleach is very toxic to fish and invertebrates.

Iodophors

Diluted at a rate of 15-20ml per litre of cold water, iodophors are also active against a wide range of fish pathogens, but are much more pleasant to use than bleach. They can even be used as a skin-scrub for personal hygiene. Use them to disinfect equipment and fishkeeping facilities as described for bleach. Iodophors are colour coded to indicate loss of activity, and should be available from pet and aquarium stores or from a veterinarian. They are supplied with full instructions for use.

Caution

- Iodophors are easily deactivated by organic matter.
- Iodophors, especially the detergent they contain, may be toxic to fish and invertebrates.

Above: Soft corals polyps such as these are among the most sensitive of aquatic organisms to the disease treatments used in the aquarium. (In fact, soft corals are difficult to maintain successfully in an aquarium at the best of times.) Since marine invertebrates are generally more vulnerable than fish to the effects of chemical remedies, it is usually better either to set up a marine fish-only system or a marine invertebrate system with a small number of relatively hardy fish. Using UV irradiation or ozone correctly and carefully can be an effective way of keeping certain disease organisms at bay and is widely applied for marines.

OZONE AND ULTRAVIOLET IRRADIATION

Ozone and/or ultraviolet (UV) irradiation may be used to remove pathogens from water containing fish, thus reducing the likelihood of disease transmission between fish or individual aquariums. As mentioned below, the use of ozone and UV irradiation can bring a number of other benefits, too. However, neither method should be looked upon as a replacement for additional disease treatments, since both only act against pathogens which are *in the water*, and not usually against those which are *attached to the fish*.

Ozone

Ozone (O_3) is an unstable form of oxygen (O_2) with powerful

oxidizing properties. As a result, it can not only bring about a fall in the number of free-swimming microbial organisms, especially viruses and bacteria, but also a decrease in the levels of organic materials that would otherwise be difficult to remove by normal filtration methods. Using ozone can eliminate the 'yellowing' of old aquarium water, for example, and also improve the efficiency of protein skimmers in marine aquariums.

Ozone generators for aquarium use are available from specialist aquatic shops. They are usually of the type that produce ozone by electrical means (by passing air through an electrical discharge), and are available in a range of sizes to produce different amounts of ozone. The ozone is usually passed into the water via an airstone. Because of its potentially toxic effects on fish, aquatic plants and invertebrates, ozone is usually employed in a separate ozonation tank or chamber into which water is pumped for treatment, perhaps as part of a total, out-of-tank filter system, or in place of air in a protein skimmer. Vigorous aeration or passing over activated carbon can eliminate ozone from treated water, before its return.

It is difficult to measure ozone concentration in water. *As a rough guide only*, beneficial effects should be achieved when ozone is applied at a rate of 0.25-1mg per hour per 10 litres of water to be treated. (Estimate this by comparing the output rating of the ozone generator to the volume of your tank.) However, the balance between providing sufficient ozone to be beneficial, yet not so much that it is toxic to the animals and plants in the system, will depend on a number of factors, including the precise method of ozone introduction and the ozone demand of the water caused by dissolved and particulate organic matter. Clearly, more ozone can be safely applied when a system is relatively densely stocked with large amounts of organic matter present, as compared to a newly set-up quarantine system containing only one or two fish.

It is known that continuous exposure to ozone, even at low levels, may be toxic to aquatic animals and plants. However, fish may be able to tolerate higher levels for relatively short periods. As a result, it may be best to use ozone in connection with a home aquarium for just a few hours each day, and in response to particular problems, rather than continuously. If at any time the fish appear distressed, irritated and/or suffering from gill problems, and particularly if there appears to be a build-up in the characteristic bleachy smell of ozone, discontinue ozonation and seek advice.

Caution

- Ozone can be toxic to humans. Its characteristic smell can be detected at 0.02-0.05mg/litre, the upper level of which may be too high for long-term human exposure. As a result, always use ozone in a well-ventilated area, and investigate any build-up in the smell of ozone, after first turning off the ozone generator.
- Even very low ozone concentrations can be toxic to fish, aquatic plants and invertebrates, including the helpful bacteria in a biological filter.
- Do not use ozone with rubber tubing, which it attacks and destroys, and ozone may also form toxic byproducts in sea water under some circumstances.
- Ozone may reduce the effectiveness of some disease treatments.

● For optimal effects, air is best dried before delivery to many of the available ozone generators.

Ultraviolet sterilization

This is another useful technique for controlling the free-swimming stages of a number of important microbial fish pathogens. In terms of its wavelength, ultraviolet (UV) falls somewhere between visible light and X-rays. UV rays generally kill microorganisms by attacking their cellular contents.

Compact UV sterilizing units are available from specialist aquarium shops, and in a range of sizes to treat aquariums and ponds of particular volumes at a given flow rate. They are usually rated on the assumption that the volume of the pond or aquarium passes through the sterilizer two to four times per hour, and are primarily active against viruses, bacteria, fungus and small protozoans. Larger organisms, such as flukes and crustaceans, need considerably higher dose rates of UV irradiation and hence UV is seldom used to combat these organisms.

Some success has also been achieved in using higher turnover UV sterilization units to control 'green water' algal problems in garden ponds. Such units are of different construction and generally operate with a higher flow rate than those used in disease control, and may not be so effective against fish pathogens.

Caution

● The tubes of an aquarium UV sterilizer produce rays which can be dangerous to the human eye. Do not look directly at an unprotected tube that is switched on.
● The effectiveness of a UV sterilizer will depend on the amount of colour and turbidity in the water. Pre-filtering or pre-treatment with efficient mechanical filtration, activated carbon or protein skimming is usually beneficial for disease control.
● UV tubes have a limited life; renew them every 6-12 months.
● Passing water through a UV sterilizer at a flow rate that is significantly above *or* below the recommended level, or attempting to use a unit in an aquarium or pond that is larger than recommended, will bring about a loss in effectiveness of the unit.

Vaccination methods

Many aquarium fish are too small to be vaccinated by injection, so alternative delivery methods are required. Incorporation of the vaccine with the food is one possibility. Studies using food-fish species have shown that certain vaccines can be delivered by live food organisms, such as brine shrimp (*Artemia*). The shrimp are fed tiny capsules of vaccine, and are themselves then fed to the fish. Bath vaccination, by immersing the

Below: A UV sterilizer unit with the outer casing removed (purely for photography), showing the tube and water jacket. (Do not look directly at an exposed UV tube).

USING TAPWATER CONDITIONERS

A variety of tapwater conditioners are available from aquarium shops. These often perform a number of useful functions, including:
● Removing chlorine, chloramine and heavy metals
● Preventing marked shifts in pH value in fresh water
● Protecting delicate and/or damaged tissues

These benefits can be of great importance when dealing with recently imported and/or damaged fish, and some proprietary tapwater conditioners have been developed to enhance the effects of certain disease treatments. However, the activity of certain disease treatments, particularly copper-based ones, may be reduced in the presence of some tapwater conditioners. Where any doubt exists, seek the manufacturer's advice. Where necessary, the active fish in a solution of vaccine for a brief period of time, has also proven successful. Another method involves temporarily removing the fish from water and spraying its body surface with diluted vaccine.

Below: For large fish, such as this ghost koi, vaccine delivery by injection is generally the most effective route.

ingredients of most tapwater conditioners can be removed by filtration over activated carbon and/or making partial water changes with unconditioned water.

If tapwater conditioners are likely to reduce the efficiency of a disease treatment, all new water may be 'conditioned' using one or more of the following methods:

- By filtration over activated carbon
- By vigorous aeration for 12 hours
- By using sodium thiosulphite (1-2 drops of 10 percent solution per 10 litres of water is usually sufficient)

While these methods do not have all the benefits of a good quality tapwater conditioner, each will remove chlorine and generally render the water safe for fish without interfering with any treatment.

Under normal circumstances, however, be sure to treat all new water with a proprietary conditioner before aquarium use.

VACCINATION AS A DISEASE PREVENTION MEASURE

Much research is being directed at the disease organisms that affect farmed fish, especially those pathogens which affect fish for eventual human consumption. In a world hungry for protein, and with dwindling catches of fish from the oceans, efficient aquacultural practices are likely to become more and more important in the future.

Since fish, just like any other vertebrate animal, can develop an immune response towards many of their pathogens, one avenue of research is the development of effective vaccines and large-scale vaccination methods to combat the viral and bacterial diseases of economic importance. In fact, commercial vaccines are already available to prevent certain bacterial diseases of salmon and trout. These include vaccines against furunculosis (caused by *Aeromonas salmonicida* bacteria), vibriosis (*Vibrio*) and enteric redmouth disease (*Yersinia*). Of more relevance to ornamental fish is the feasibility of vaccinating koi and goldfish against ulcer disease – an increasingly common disease caused by various aeromonad bacteria and especially atypical strains of *Aeromonas salmonicida* – using one of the commercial furunculosis vaccines. The ongoing development of an SVCv (Spring Viraemia of Carp virus) vaccine could bring future benefits to koi keepers by protecting their fish against this serious and highly infectious disease. Research into vaccines against certain fish parasites, notably the white spot parasites *Ichthyophthirius* and *Cryptocaryon*, is also underway.

The conventional and generally most effective method of vaccinating fish is by injection; however this is not suitable for many tropical aquarium fish because of their small size. Fortunately, alternative vaccine delivery methods are possible, such as via the water (immersion vaccination) or orally by incorporating the vaccine with the feed.

In the future it may even be possible to purchase aquarium and pond fish that have been vaccinated against certain infectious diseases.

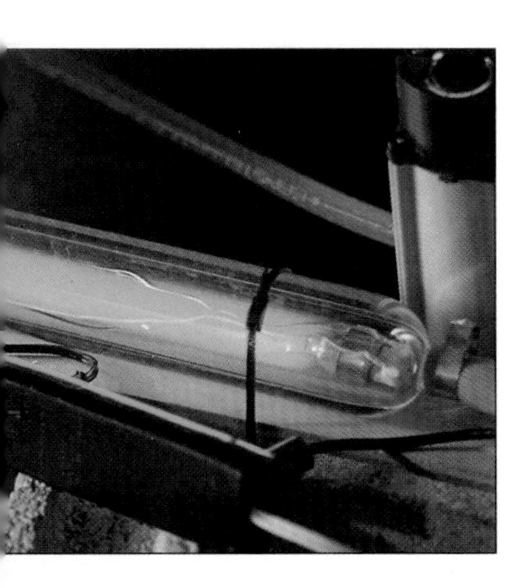

INDEX

D

E

F

G

H

I

K

L

M

N

O

P

PICTURE CREDITS

Artists

Copyright of the artwork illustrations on the pages following the artists' names is the property of Interpet Publishing. The artwork illustrations have been credited by page number.

Rod Ferring: 12-13, 26, 27, 28, 29, 35, 38, 42-3, 50-1, 52, 53, 59, 61, 72-3, 82-3, 84-5, 86-7, 88-9, 90-1, 92-3, 97, 98, 113, 114, 117, 119, 121, 125, 127, 129, 149, 157, 160, 168, 171, 173, 175, 176, 180

Bill Le Fever: 18, 18-19, 22, 25, 34, 47, 51, 79, 80, 81

Photographs

The publishers wish to thank the following photographers and agencies who have supplied photographs for this book. The photographs have been credited by page number and position on the page: (B) Bottom, (T) Top, (C) Centre, (BL) bottom left etc.

David Allison: 16-17(TC), 19, 20-1(B), 22-3(B), 26-7(B), 30-1, 32-3, 40(C), 85(TR), 93(TL,C), 122, 140, 141(B), 152-3

Chris Andrews: 13(T), 29, 64-5(B), 76, 81, 82(C), 83(TR), 83(BL), 84(BR), 85(C,BR), 86(TL), 88(C), 89(TC,BL), 90(C,B), 91(TL), 92(BL), 93(B), 99(C), 100-1(C), 101(TR), 102(B), 109(TR), 111(B), 112(C), 113, 114, 115(T), 117(TR), 118(R), 120-1(T), 124(T), 126(T), 135(CR), 137, 138, 142-3, 143(B), 144(T), 147(TL), 148(L), 149, 152(L), 155(T), 156(T), 157, 161(B), 162-3(C), 166(T), 169(B), 170(T), 172(TL), 173, 177(B), 181

Peter Burgess: 79, 81, 93(TL), 109(TR), 112(C), 136(B), 162(B), 191, 194, 201(T)

James Chubb: 82(TL), 84(TR), 91(TR,C,B), 98-9, 100(B), 106(C), 106-7(T), 108-9(T), 115(B), 116-7(TC), 119, 120(B), 125, 128, 155(C,B), 160, 161(T), 170(B), 171, 172(T,B), 174-5, 179(B)

Eric Crichton © Salamander Books Ltd: 24-5(T), 38, 39(T), 56, 60-1(T), 70-1, 92(T), 132(B), 186

K A Frickhinger: 88(TR), 126(B)

Max Gibbs: 69

Les Holliday: 188-9

Andy Horton: 40-1(T)

Roger Hyde: 12(T)

Jan-Eric Larsson: 17(B), 20, 36-7(B), 74-5, 82(TR), 82(BL), 83(C), 84(BL), 88(BR), 106(B), 108(B), 152(B), 152(TR), 158, 167(T)

Dick Mills: 187

Barry Pengilley 143(T)

Laurence Perkins: 56-7(T), 112(T)

Geoff Rogers © Interpet: 12, 24, 38, 39(T), 44, 45(T), 48, 57(B), 58(B), 60(B), 63(B), 64(T), 65(T), 189

Fred Rosenzweig: 12-13(B), 58-9(T), 88(TL,BL), 94-5, 107, 111(T), 151, 154, 164-5, 167(B)

Salamander Books Ltd: 44-5, 48, 57, 58, 60, 63, 64

Mike Sandford: 16(B), 17(TR), 28, 35, 39(B), 49, 50-1, 54-5(T), 86(B), 87(B), 89(C), 92(BR), 93(TR), 99, 103, 105(B), 110(B), 112(BL), 122-3(C), 130-1(TC), 134(B), 142(B), 145(T), 200

David Sands: 40-1(B), 83(TL), 84(C), 124(B), 147(TR), 162, 168, 178

Peter W Scott: 46-7(T), 55, 82(BR), 83(BR), 85(BL), 86(TR), 87(TR), 88(TC), 89(TR), 96(B), 97(T), 102(T), 105(T), 110(TR), 111(C), 116, 118(L), 123(TR), 130(B), 144(B), 147(B), 148(T), 148-9, 150, 156(B), 163, 169(T), 179(T), 201

W A Tomey: 3, 10-11, 14-15, 23, 26-7(T), 37, 42-3, 46(B), 62-3(T), 66-7, 84(TL), 85(TL), 87(TL), 89(TL,BR), 90(T), 92(C), 96-7(TC), 104-5(B), 109(B), 132-3(T), 134-5(T), 135(BC), 136, 136-7, 138-9(T), 139, 140-1, 141(T), 145(B), 146(B), 159, 166(B), 177(T), 182-3

Chris Williams: 83(BL), 87(T), 96(BR), 98(T), 99(CL), 103, 126(T), 145(B), 154(BL), 169(B), 174(B), 181(T), 181(B)

William H. Wildgoose: 151(T), 155(C)

FURTHER READING

Bassleer, G. *Diseases in marine aquarium fish*. Bassleer Biofish (distributor), 1996.

Burgess, P. and Bailey, M. *A-Z of tropical fish diseases and health problems*. Ringpress Books Ltd., 1998.

Hoole, D., Bucke, D., Burgess, P. and Wellby, I. *Diseases of carp and other cyprinid fishes*. Blackwell Science Ltd. (Fishing News Books), 2001.

Jepson, L. *Koi medicine*. TFH Kingdom Books, 2001.

Untergasser, D. *Discus health*. TFH Publications, Inc., 1991.

Untergasser, D. *Handbook of fish diseases*. TFH Publications, Inc., 1989.

Advanced texts on fish health

(The following titles are written primarily for fish health scientists and veterinarians)

Lewbart, G.A. *Self-assessment colour review of ornamental fish*. Manson Publishing/The Veterinary Press, 1998.

Lom, J. and Dykova, I. *Protozoan parasites of fishes*. Elsevier Science Publishers B.V., 1992.

Noga, E.J. *Fish disease: diagnosis and treatment*. Iowa State University Press, 2000.

Ross, L.G. and Ross, B. *Anaesthetic and sedative techniques for aquatic animals*. Second edition. Blackwell Science Ltd., 1999.

Wildgoose, W.H. (editor). *Manual of ornamental fish*. Second edition. BSAVA Publications, 2001.

Hobbyist magazines

Many aquarium and pond-fish magazines carry regular features on fish health. Examples are: *Practical Fishkeeping* (published in the UK); *Today's Fishkeeper* (formerly *Aquarist and Pondkeeper*)(UK); *Koi Ponds and Gardens* (UK); *Tropical Fish Hobbyist* (USA); *Freshwater and Marine Aquarium* (USA).